Principles of Muscle Biology for Meat Science

PRINCIPLES OF

MUSCLE BIOLOGY

FOR MEAT SCIENCE

Seon-Tea Joo

GNU PRESS

This book is dedicated to the memory of my graduate student, Sang-Hoon Moon, a true disciple of mine and the best student in Meat Science Laboratory at Gyeongsang National University. May he rest in peace.

Preface

As animal muscle is the most valuable source for nutritive and tasteful foods, various efforts have been made in the world over few centuries to understand the relationship between muscle and meat. Because muscle is the primary component of meat, an understanding of muscle biology is basic to meat science and meat industry practices. Especially, a knowledge of muscle structure is critical to understanding the interrelationships between properties of muscle and its use as meat. Also, it is essential to understand the mechanism of muscle growth and development for improvement of eating quality and economical values. Furthermore, it is definitely necessary to understand a mechanism of muscle contraction because muscle shortening decreases meat quality during postmortem muscle conversion to meat. In this regard, this book provides a broad-based coverage of the muscle structure, growth and development, contraction mechanism, and conversion muscle to meat.

The primary goal of this book is to provide a text for not only undergraduate but also graduate students enrolled in meat science and animal science courses. I believe that this text is suitable for all students enrolled in their first meat science course. Also, I wish that this book will be helpful for not only students but meat scientists and the people who work in the meat industry.

I would like to express my specific indebtedness to Ishamri Ismail, my Ph.D. student, for his assistance to write this book. Also, I express my thanks to all members of Meat Science Laboratory, Department of Animal Science, Gyeongsang National University, Jinju, South Korea.

Seon-Tea Joo

Contents

Chapter 1.

Structure of Muscle

A knowledge of muscle structure is critical to understanding the interrelationships between properties of muscle and its use as meat. Meat consists of large muscle and variable portions of connective tissues. It incorporates additionally a few epithelial and nervous tissues. The cells of epithelial tissue are generally applied closely to one another with little cementing substance. Sheets of cells may develop into the covering of the outer and inner surfaces of the body, and the glands and other structures are derived from them. The cells of nervous tissue are concerned primarily with the rapid conduction of impulses in the integration of numerous functions. Muscular tissue is associated with movement and position of the skeleton and contractility in many organs, including the vascular system. Many connective tissue types which include adipose tissue, bone, cartilage, and connective tissue are also present in the carcass. In connective tissues the cells are generally separated from one another to a greater or lesser degree by intercellular substance. The blood cells and blood-forming tissues and cartilage and bone are derived from the connective tissue cells.

Before further discussions towards muscle and associated tissues, a number of anatomical terms must be understood to describe the relative positions of structures within the body. Common terms such as front, back and sides are of little use because of the changes in posture and limb position that occur when a standing animal becomes a hanging carcass.

Anterior	– towards the head
Posterior	– towards the tail
Dorsal	– towards the upper part or back of the standing animal
Ventral	– towards the lower part or belly of the standing animal
Medial	– towards the midline plane that separates right and left sides of the body
Lateral	– towards the sides of a standing animal
Proximal	– towards the body in a limb of the animal
Distal	– away from the body in a limb of the animal

1. Muscle Tissue

In general, there are three types of muscle: (1) striated voluntary or skeletal; (2) striated involuntary or cardiac muscle; and (3) smooth or involuntary muscle. Skeletal and cardiac muscles are referred to striated muscles because of the transverse banding pattern observed microscopically. Smooth muscle fibers are long, spindle-shaped cells, which are thick in the middle and have narrow tapered ends.

1.1. Skeletal Muscle

Skeletal muscles are composed of bundles of muscle fibers embedded in connective tissue and some of them are connected to bone via ligaments, fascia, cartilage or skin. They contain the other tissue types sorts along with connective tissue, nerve and epithelia. These features of muscle are shown in Figure 1.1.

The individual muscle cells, muscle fibers, are encased in collagenous connective tissue that is called the endomysium, which forms a continuous connective tissue network from the muscle cell surface to the epimysium. Muscle cells are grouped into bundles that are surrounded by a layer of connective tissue called the perimysium. Small bundles are grouped into larger bundles, which eventually form the muscle (Figure 1.2). Each muscle bundle contains about 50~150 muscle fibers. Most muscles attach to bones. The muscle connective tissue (type III collagen) become continuous with the tendon connective tissue. Tendons contain type I collagen and are lubricated to facilitate the gliding action of the tendon.

The connective tissue provides a supporting frame (skeleton) to connect and hold different parts of the body. It consists of bones,

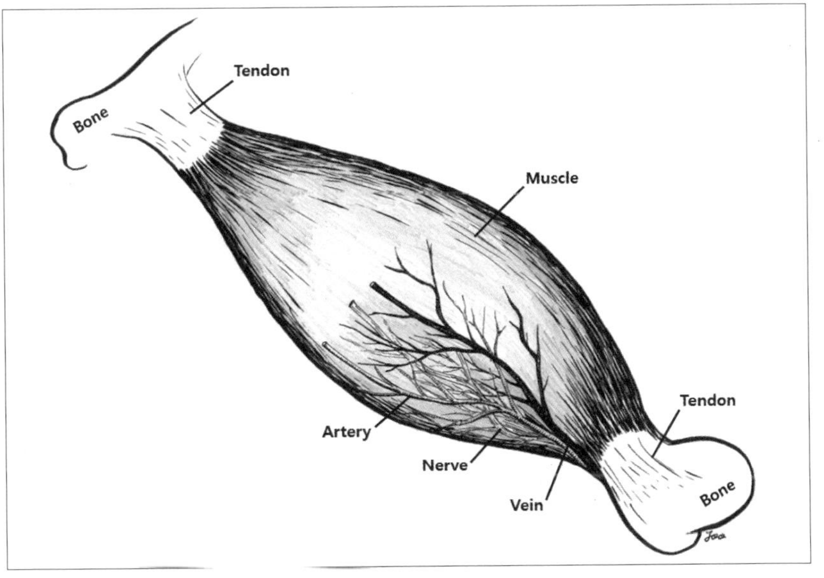

Figure 1.1. Drawing of a skeletal muscle showing the structural relationships among tendons, blood vessels, and nerves.

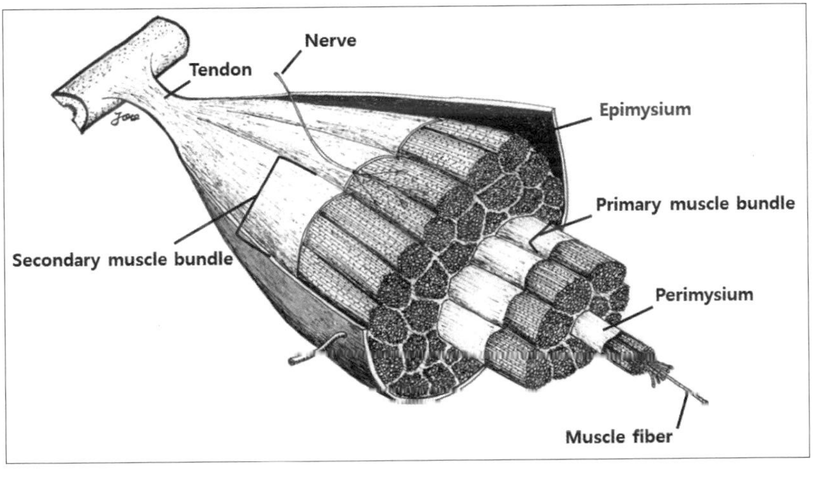

Figure 1.2. A diagrammatic illustration of the structure of skeletal muscle

ligaments, connective tissue covering muscle bundles and fibers, adipose tissue and blood. The tissue responsible for building bones and cartilage is called "supportive connective tissue" because it provides strong structural support. Tissue that surrounds muscles, muscle bundles and fibers is called "connective tissue proper". The two types of supportive connective tissue show a number of similarities in their composition and functionality. Usually, both consist of few cells and a lot of extracellular substance. The tissue can range from very soft to very tough such as bones that contain embedded fibers and mineral crystal (calcium salts). In bones, the extracellular substance is tougher than in other connective tissues such as cartilage, where the extracellular substance is more rubbery and soft.

Skeletal striated muscle fibers are multinucleated, elongated and more or less tubular cells. The ends may be conical or tapering, and the diameter of the fiber ranges 10-100 μm. Fiber length varies greatly, ranging from 1-40 mm, but in most cases the fiber probably does not run the entire length of the muscle. In some muscles, the fibers may run in a longitudinal direction to the long axis of the muscle, but the usual case is much more complicated arrangement. The muscle fiber is encased in the sarcolemma, which is delicate membrane immediately under the endomysium. This relationship is illustrated in Figure 1.3. The characteristic cross-striated appearance of the fiber is due to the presence in the fiber of a series of thin cross-striated unit known as myofibrils. The muscle fiber also consists of intracellular membrane

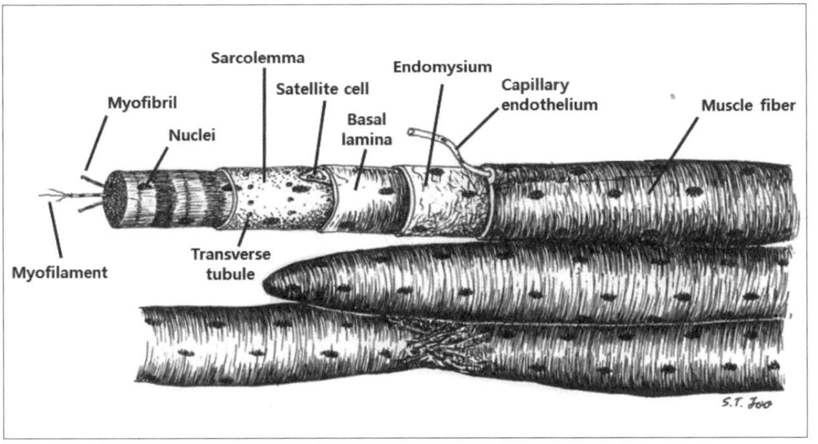

Figure 1.3. Drawing of skeletal muscle fibers showing structural features and their longitudinal orientation.

network that include the T-tubules and sarcoplasmic recticulum, as well as mithochondria and nuclei (Figure 1.4).

Myofibrils are bathed in sarcoplasm and extend the entire length of the muscle fiber. Muscle fibers of meat animals with diameters of 50 μm have at least 1000 and could have over 2000 myofibrils. Myofibrils are known as contractile or cytoskeletal proteins. The main proteins are myosin and actin, which make up the thick and thin filaments, respectively. The arrangement of these two myofilaments in myofibrils, and the fact that thick and thin filaments overlap in certain regions along their longitudinal axes, account for the striated appearance of the myofibrils and muscle fibers of skeletal muscle. More detail ultrastructure of muscle fiber is described at Chapter 2. Overall, this

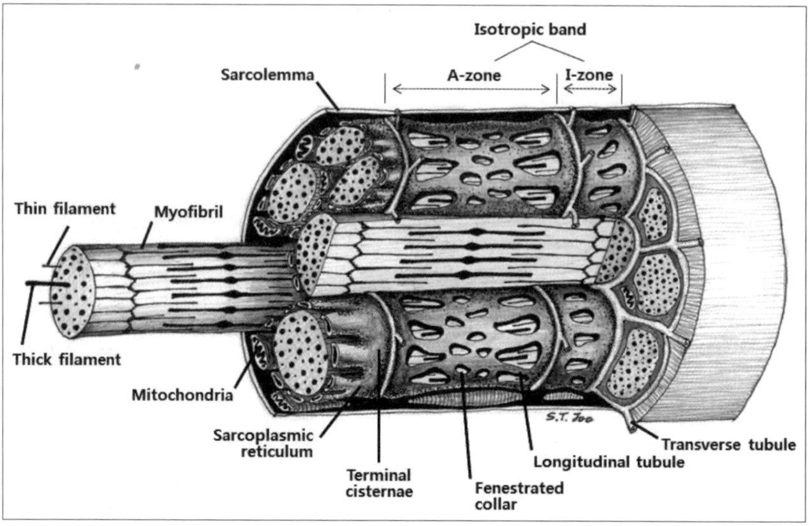

Figure 1.4. Diagrammatic representation of the sarcoplasmic reticulum and T tubules, and their relation to the myofibrils of mammalian skeletal muscle.

myofibrils represents about 55% of the muscle proteins.

Nervous tissue serves as the communication system within the body. While it represents a small part of the edible meat (usually less than 1%). Nerve fibers and blood vessels enter and exit the muscle along these connective tissue networks (Figure 1.5). Each muscle is supplied with at least one artery and vein, as well as with nerve fibers from a nerve trunk. Nerve fibers and blood vessels supplying the muscle are associated with the epimysium and they enter its interior via the perimysium. In turn, branches that arise from these nerves and blood vessels supply individual muscle fibers and are supported by the endomysium.

A tendon is a projection of connective tissue (type III collagen) beyond the ends of the muscle that attaches to bone. Tendons contain type I collagen and are lubricated to facilitate the gliding action of the tendon. In the electron micrographs of the myo-tendon junction, the end of the muscle fiber shows many processes and invaginations, which correspond to the zig-zag appearance in the light micrographs. As the myofibrils approach the tapered ends of the muscle fiber, the characteristics banding pattern is lost and the myofibrils become continuous with strands of noncontractile fibers. Tendon fibrils are inserted in the sarcolemmal membrane, opposite the point at which noncontractile connecting fibers of each myofibril are attached. This entire myotendinal structure is continuous with the tendon (Figure 1.6).

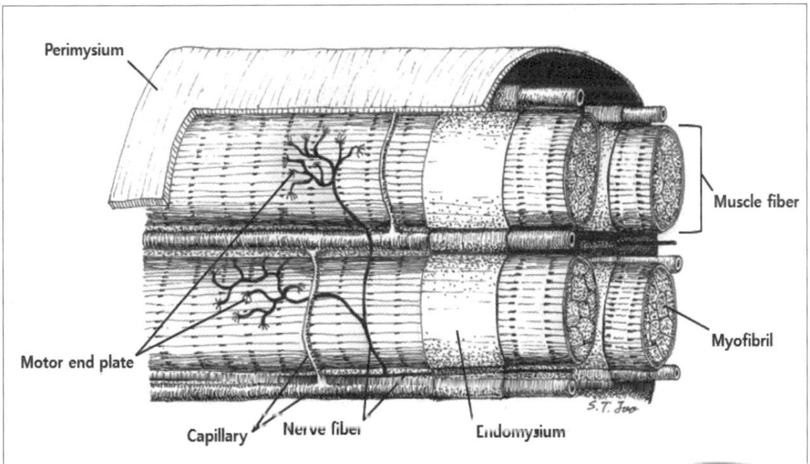

Figure 1.5. Diagrammatic illustration of the structure of a primary muscle bundle, showing the relationship of fibers, endomysium, blood vessels, and nerves.

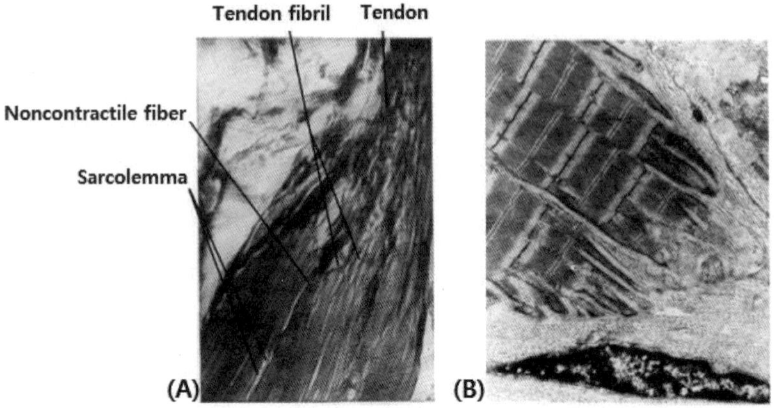

Figure 1.6. Photomicrographs of the muscle-tendon junction. (A) Visible light photomicrograph showing a muscle-tendon connection (×500). (B) An electron photomicrograph of the myotendinal junction, showing the relation of the sarcomere to tendon collagen fibers.

The collagen fibrils, which leave the surface of the end of the muscle fiber, run for some distance and join the main bundles of the tendon. In other words, there are small bundles of collagen fibrils connecting between the end of the muscle fiber and the typical tendon tissue. Several fibroblasts showing somewhat different appearance from typical tendon cells are always seen at this region (Figure 1.7).

1.2. Cardiac Muscle

Cardiac muscle is found mainly in the heart and in large blood vessels where it serves as a pump to generate the pressure needed to transport

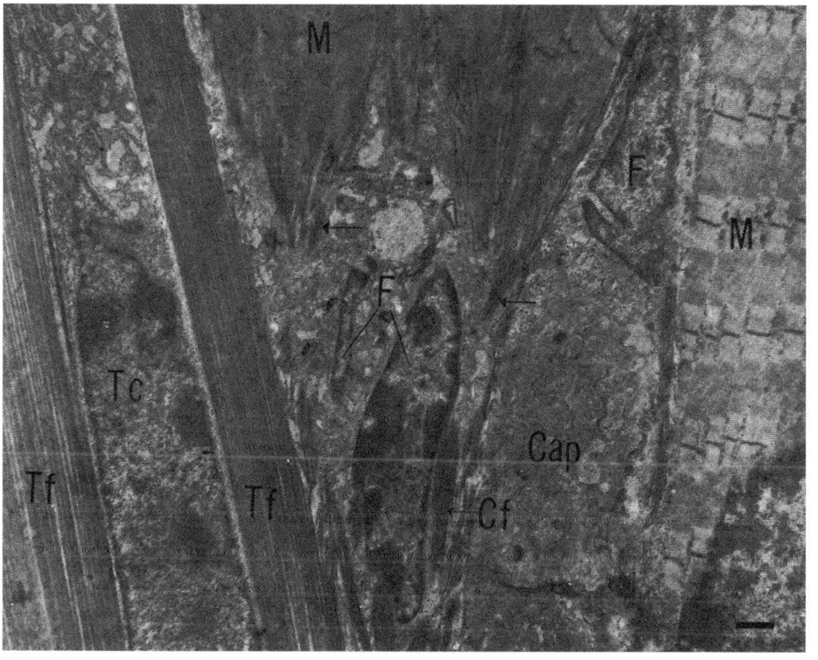

Figure 1.7. Electron micrograph of a myo-tendon junction. Collagen fibrils (Cf) at the myo-tendon junction obliquely join the main bundles of the typical collagen fibrils (Tf). M muscle fiber, F fibroblast, Cap capillary, Tc tendon cell. Young adult guinea-pig (5800x)

blood through the circulatory system, a most vital process. Cardiac muscle is an involuntary muscle over which the animal has no direct control. The cell have a striated appearance like a skeletal muscle, but have only one or two nuclei per cell and have a dark red color as a result of its extensive blood supply. The average length of the cell is about 50 100 μm, and its width is about 15 μm. Cardiac muscle has a unique rhythmic contraction that is triggered by the sinoatrial node and that starts early on in embryonic development. The heart is controlled by the sympathetic and parasympathetic nervous systems, which are partly outside the control nervous system.

Another unique structural characteristic of cardiac muscle is that the fibers run in a mesh-line pattern and are branched. This allows the heart chambers to contract (reduce volume), and pump blood forward. Microscopic examination reveals unique structures called intercalated disks that appears as dense lines, at regular intervals, along the longitudinal axis of cardiac fiber. They provide a cohesive link between the fibers and facilitate the transmission of contraction force from one fiber to the other.

The myofilaments of cardiac muscle are not well organized as like skeletal muscle, although thick and thin filaments in myofibrils are aligned to give a striated appearance. The myofibrils are discrete and dimensions vary along the longitudinal axis of the fibrils. Consequently, myofibrils are extremely variable in size. Elements of the sarcoplasmic reticulum are less developed and the T tubules are larger in diameter compared to skeletal muscle. T tubules occur at the Z disks, in contrast to their occurrence at the A-I band junction in skeletal muscle. Also, structures comparable to the terminal cisternae are absent (Figure 1.8).

Cardiac muscle fibers give off branches and so effect anastomoses with adjacent fibers. In cross-section, the fibers appear rounded or irregular, display centrally placed nuclei. The fibers are packed with myofibrils which are similar to those of striated skeletal muscle.

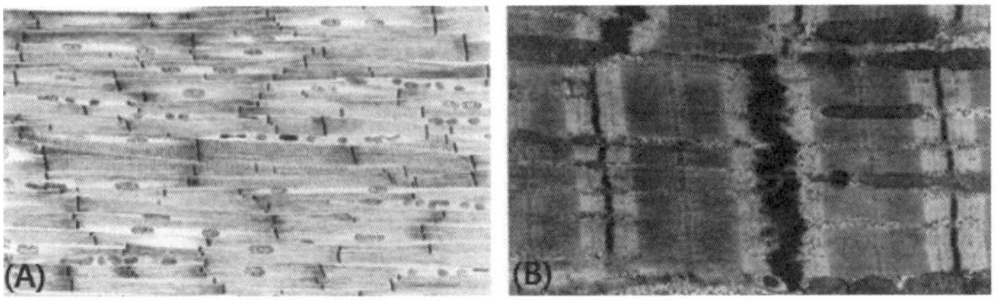

Figure 1.8. Photomicrographs of cardiac muscle. A longitudinal section (A), showing fiber arrangements and nuclei locations and, (B) an electron photomicrograph showing the banding pattern, intercalated disks, mitochondria, and glycogen granules.

Sarcoplasm is abundant and contains more numerous mitochondria than skeletal striated muscle. The intercalated disc is the distinguishing feature of cardiac muscle. It is a dense area oriented transverse to the long axis of the fiber. Electron microscopy has clearly revealed that the myocardium is made up of individual cellular units and that intercalated disks are the sites of their end-to-end attachment.

The myocardium is the contractile layer of the heart and contains the bulk of the cardiac muscle. Fibers of the myocardium are held in place by connective tissue fibers, which are continuous with connective tissue sheaths that group them into fascicles (bundles) of fibers. Blood and lymph vessels and nerve fibers enter and exit the myocardium via the connective tissue between muscle bundles. Cardiac muscle contains numerous glycogen granules in sarcoplasm. Also, the mitochondria of cardiac muscle are especially large and numerous. Cardiac muscle is endowed principally with a capacity for oxidative metabolism; consequently, it has a more extensive capillary network than either smooth or skeletal muscles.

1.3. Smooth Muscle

Smooth muscle comprises only a small proportion of meat and are part of the involuntary systems in the body (i.e. the digestive system, the walls of arteries, and parts of the reproductive system). The fibers have a single, centrally located nucleus and are relatively long and narrow with an average length of a few hundred μm and a diameter of 3-12 μm. This muscle does not have a striated appearance like skeletal and cardiac muscle because the repetitive structure of the sarcomere is not as well organized, hence the name smooth muscle. In terms of layout within the body, some areas show different layers of smooth muscle. For example, in the digestive system, a cross section reveals smooth muscle layers that are positioned both perpendicular and parallel to the cut surface. This allows the digestive system to both decrease the gut diameter and elongate to move food down the tube-like structure.

However, whatever the arrangement, each fiber or bundle is surrounded by a delicate network of fine connective tissue that support and bind them in place. This connective tissue includes collagenous and elastin filaments, and an associated ground substance. Also, blood vessels and nerve fibers are associated with this connective tissue.

Although this connective tissue matrix occupy the narrow space between the muscle fibers, smooth muscle is poorly supplied with blood compared to skeletal muscle.

Smooth muscle cells have one nucleus per cell and are long spindle-shaped cells, having lengths up to 0.5 mm. The cells are often found in bundles or sheets. Contraction of smooth muscle can be initiated by nerve impulses, hormonal stimulation, or changes in the muscle cell such as stretching. The major contractile proteins in smooth muscle are myosin, actin, tropomyosin, α-actinin, and filamin; however, troponin is not present. As the name implies, nonstriated muscle lacks the filamentous organization found in cardiac and skeletal muscle. However, smooth muscle fibers vary in size and shape, depending on their location. They are not always spindle shaped, as usually depicted, but may be tremendously uneven in contour along their length. In cross-section, smooth muscle fiber vary from extremely flattened ellipsoids to triangular and polyhedral shapes (Figure 1.9).

Smooth muscle cytoplasm often appears homogeneous but does contain fine fibrils which run the length of the cell. The fibrils are homogeneous and lack alternating dark and light bands, which are characteristic of striated muscle myofibrils. While striated skeletal muscle adjusts the organism to the external environment, smooth muscle helps maintain the internal environment or physiological balance of the body. Smooth muscle is found in such areas as the viscera, blood vessels, lymphatics, and skin, and it is usually closely

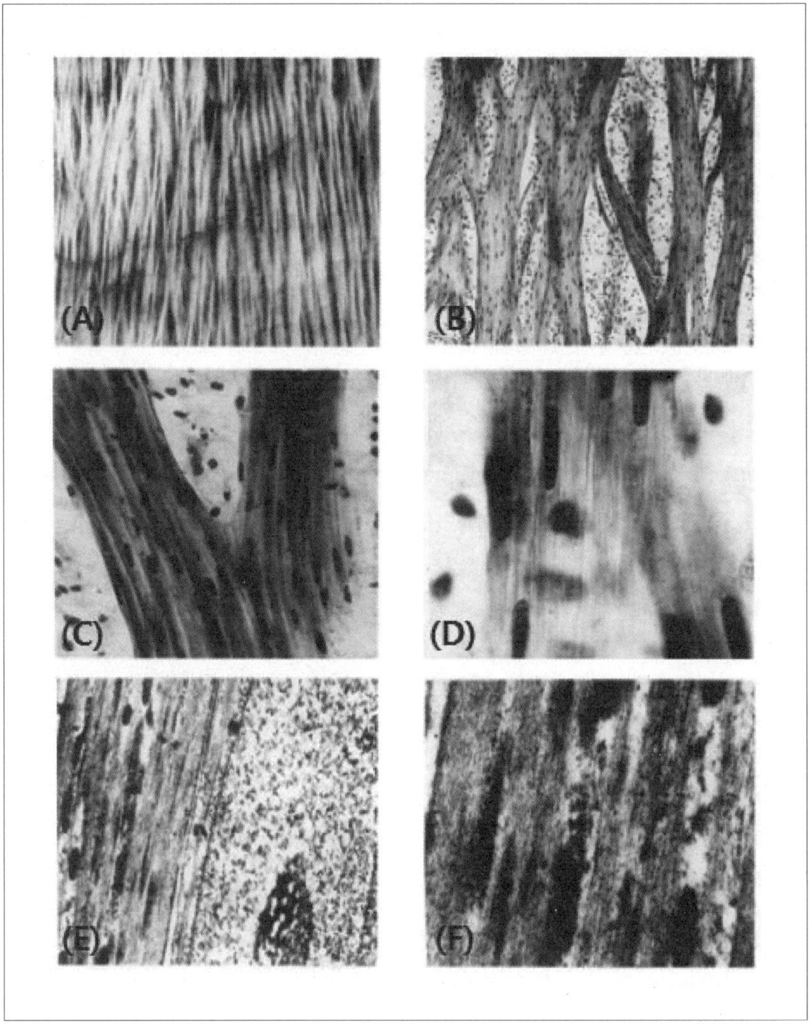

Figure 1.9. Photomicropraphs of smooth muscle fibers from the longitudinal muscle layer of a pregnant rabbit's uterus at term. (A-D) light microscopy (×12, ×75, ×300, and ×1000, respectively) and, (E-F) electron photomicrographs (×10,000 and ×20,000, respectively).

associated with connective tissue layers or strands. Because of the arrangement of smooth muscle in walls of vessels to conduct peristaltic movement, it often appears in section as a circular layer around the vessel adjacent to a longitudinal layer running the length of the vessel.

Myofilaments of smooth muscle fibers are also less ordered than in skeletal muscle fibers. They are not arranged in groups to form myofibrils. However, the myofilaments are arranged in pairs that run parallel to the longitudinal axis of fiber. A large number of filaments, the densest aggregation of filaments, are found at each end of smooth muscle fibers. Thick myosin filaments are not visible, even though myosin is very abundant in smooth muscle. Myosin and actin are present in the same proportions as in skeletal muscle, but there are no striation in smooth muscle.

The myofilaments of smooth muscle are attached to dark zones on the sarcolemma. These dark zone are evident along the surface of the sarcolemma, but they are less abundant than that located at the ends of fibers. These dark zones are described as being analogous to the Z disks of skeletal muscle, since they move with contractile action. These structures are responsible for transmitting the myofilamental contraction force to the sarcolemma.

2. Connective Tissue

Connective tissue connects and holds various parts of the body together. Connective tissue are distributed throughout the body as components of the skeleton, in the framework of organs and blood and lymph vessels and in sheaths that surround structures such as tendons, nerve trunks, and muscles. The skin or hide is attached to the body by connective tissues. Connective tissues provide the body with a barrier against infective agents, and are critically important in wound healing. Fat storage cells are located within a specialized type of connective tissues known as adipose tissue.

Connective tissue surrounding muscles, muscle bundles, and muscle fibers is fibrous and is known as connective tissue proper. Bone and cartilage, on the other hand, are supportive connective tissues because other tissues are attached to them, providing the body with structural support. Although preceding generalizations about connective tissues might indicate that they are quite different from each other, they have many similarities in composition and function. Fibers of connective tissue proper are sometimes continuous with bone and cartilage, and some chemical compounds are common to a number of connective tissues.

Connective tissue proper – consists of fibers with special helix structure of the collagen molecule which provides both strength and elasticity (Figure 1.10). Tropocollagen molecules are the basic structural units of the collagen fiber. They are composed of three α-chains that form a triple helix. There are about a dozen types of collagen molecules that have different functional properties and, accordingly, can be found

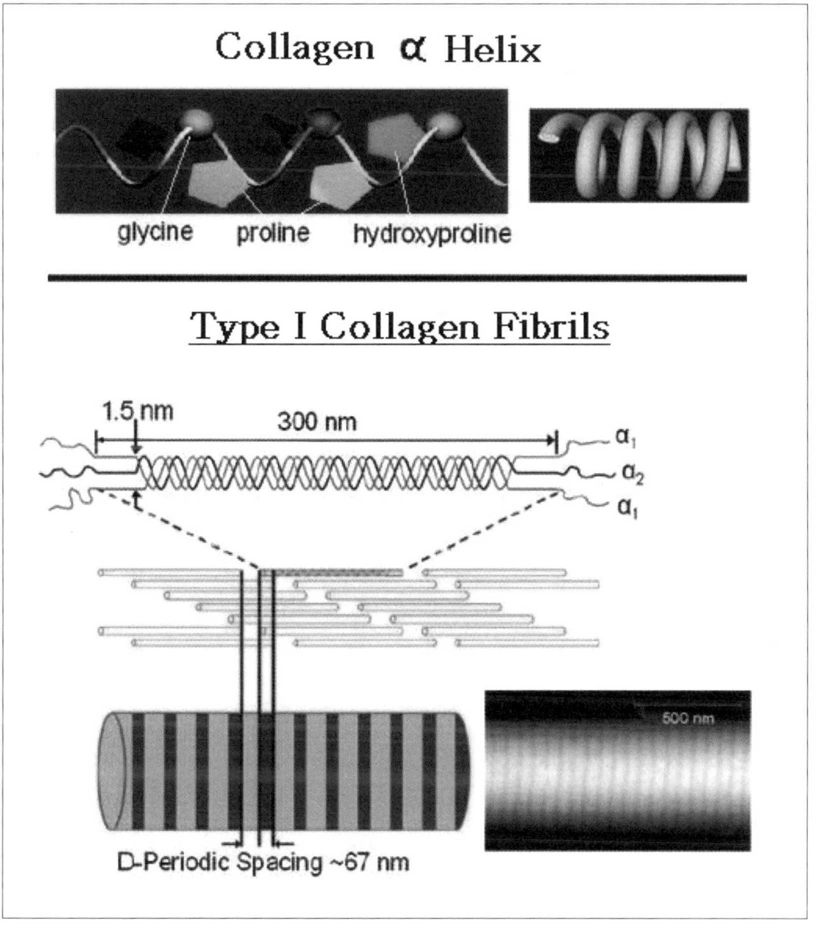

Figure 1.10. Microstructure of a collagen fibril which shows how the fibril, participating in muscle movement, can stretch. A striation pattern is seen at 64-67 nm intervals due to the parallel layout of the tropocollagen molecules. The striation can be seen after negatively staining a sample with heavy metals and viewing with an electron microscope. Note: collagen has a relative high content of the unique amino acid hydroxyproline, which can be also used to quantify the amount of collagen in muscle food.

in different locations in the body. The different types of collagen result from at least 20 different α-chains that can be combined in different ways to form the triple helix. During filament assembly, the tropocollagen molecules are aligned longitudinally, end to end, and laterally in a slightly overlapping stagger as shown in Figure 1.10. This unique spacing and overlapping of tropocollagen molecules results in a collagen fiber that has a striated appearance. Not all types of collagen form fibers. Type I and III form large and fine fibers, respectively, type IV is non-fibrous and forms a wire-like sheath that surrounds individual muscle fibers (basal lamina), and type V and VIII form microfilaments. In general, the number of collagen fibrils within a muscle depends on its expected load, stress and activity. Another factor that contributes to overall strength is the formation of intermolecular cross-linkages among the collagen fibrils. In young animal, there are few cross-linkages but as the animal ages the number increases and the bonds become more difficult to break.

Elastin is another major connective tissue protein with a different structure. As compared to collagen, it has a rubbery texture and its fibers can be easily stretched before returning to their original length. Elastin is commonly found in ligaments and arteries and provides structure to certain organs. Connective tissue proteins usually represent about 1.0% of the total muscle composition. In the meat industry the amount of connective tissue is commonly assessed by the quantity of hydroxyproline, an amino acid that is unique to collagen. Older animals are known to have tougher meat because of increased cross-linking in the collagen fibers. Both processors and consumers should know that collagen can be broken down by exposure to heat, especially prolonged, moist heat that can break some/all of the cross bridges and eventually turns collagen into gelatin. Some of the collagen becomes soluble during cooking (starting melting point ~67°C). As exposure time and temperature increase, more collagen will be converted into gelatin, which becomes apparent when the meat cools and its juices have a jelly-like consistency. Elastin, on the other hand, cannot be broken down by heat. Therefore, areas high in elastin should be either discarded or tenderized by mechanical means (needles or small blades).

Bones are also part of the connective tissue. Bone is an active tissue where building and degradation occur all the time. It consists of an organic matrix and inorganic salts. The former contains the collagen fibers and so called ground substance that consists of protein and sugar complexes. The latter is primarily made up of calcium salts (calcium phosphate and calcium carbonate), which form crystals deposited within the collagen fibers of the organic matrix. The structure consists of bone cells distributed within the matrix and arranged in small cylindrical elements called lacuna (Figure 1.11). These structures form a network of canals between the cell cavities that are important in delivering cell nutrients.

The overall structure of bone (e.g., ulna, femur) is shown in Figure 1.11 and 1.12. The long shaft, called a diaphysis, is filled with marrow while the outside consists of hard, compact bone structure consisting

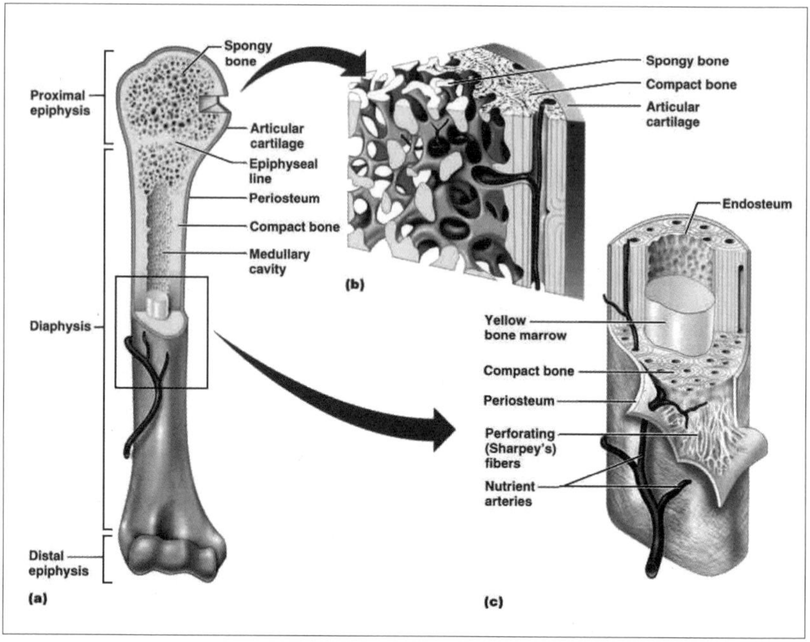

Figure 1.11. Overall structure of a long bone

of an organic matrix and inorganic salts. Both ends of the bone, called epiphyses, are enlarged to allow sufficient surface area to connect with other bones via a cartilage-mediated medium. The epiphyseal growth plate is a region underneath the epiphyseal cartilage that separates the diaphysis and epiphysis and is the region responsible for bone elongation. The central hollow part of the bone contains the bone marrow that produces new red blood cells. As indicate above, bone tissue a dynamic system in terms of calcium deposition and withdrawal. For example, the laying hen is used as a research model to study osteoporosis in humans because of the fast calcium turnaround during the laying period.

Cartilage is another connective tissue that has a strong structure and is used to connect and support different skeletal elements. Cells within cartilage are called chondrocytes and are found in cluster located in small cavities within the extracellular material. The interlacing collagen forms a delicate network of cartilage. Cartilage can differ in the relative amount of collagen fibers and extracellular material. This results in the formation of cartilage with different properties, of which there are three main categories. The first is hyalin cartilage, which is found between individual vertebrate, on the surfaces of joints and bones, and on the

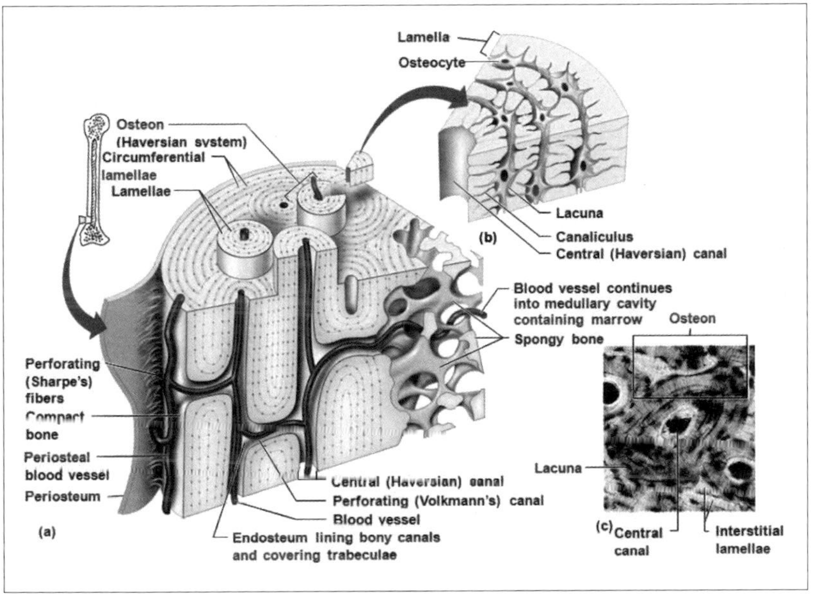

Figure 1.12. Structure of a long bone and its microstructure.

dorsal tips of vertebrae. The second type is fibrocartilage, which is found in tendons and within joint ligaments. Fibrocartilage has numerous collagen fibers and can resist repetitive stress. The third type is elastic collagen, which consists of a number of branched elastin fibers that provide elastic characteristics.

Adipose tissue consists mainly of cells and functions to protect sensitive organs (cushioning), store fat (energy) and insulate parts of the body. Adipose tissue is the main means of energy storage for the animal and is used in response to certain needs. Adipose tissue is usually found enclosed in areas surrounded by a sheath of collagen fibers. Young adipose cells are called adipoblasts. After they mature and fill with fat, however, they are called adipocytes. Adipoblasts grow from 1-2 μm to size of up to 100 μm by accumulating small lipid droplets that fuse to form a large fat globule. Adipose tissue development is related to age of the animal and the amount of available nutrients. In young animals, the first fat deposit usually appears in the visceral area. Later, subcutaneous fat (under the skin) is developed, followed by a limited amount of intermuscular fat that is deposited in between muscles. In any case, the adipose tissue has a fairly dynamic metabolism, meaning that stored lipids are constantly mobilized.

3. Nervous Tissue

Nervous tissue constitutes a small proportion of meat (less than 1 percent), but functions in the period immediately prior to and during slaughter influence on meat quality. Understanding its structure is essential to understanding muscle contraction, post-mortem changes, and meat quality issues. The two main structural components are the central nervous system (brain and the spinal cord) and the peripheral system that consists of the nerve cells that reach all parts of the body.

The nerve cell, or neuron, is the basic building block of nervous tissue and has a distinct structure consisting of a cell body with an elongated fiber-type structure called an axon (Figure 1.13). A nucleus is found within the polyhedrally shaped cell body. A few short branched structures called dendrites come out from the body. Motor neurons are those that reach muscle fiber and have a long, single axon that branches when it reaches the muscle. The junction points are called the motor end plates, or neuromuscular junctions (i.e. one nerve reaches a number of muscle fibers and triggers them all concurrently). The action potential, an electric pulse of about 80 mV that goes from the cell body through the axon to the motor end plates, is transferred to the muscle or other nerves (i.e. their dendrite portion) via a synapse, which is a physical gap between the two cells or structures. Therefore, a chemical transmitter called acetylcholine is used, in most synapses, to convey the message across the gap (e.g. to the muscle). It should be noted that within the brain and other locations, other chemical messengers are used. Certain toxins can block acetylcholine and cause serious problems to the animal. An example of a serious toxin that is of importance in the food

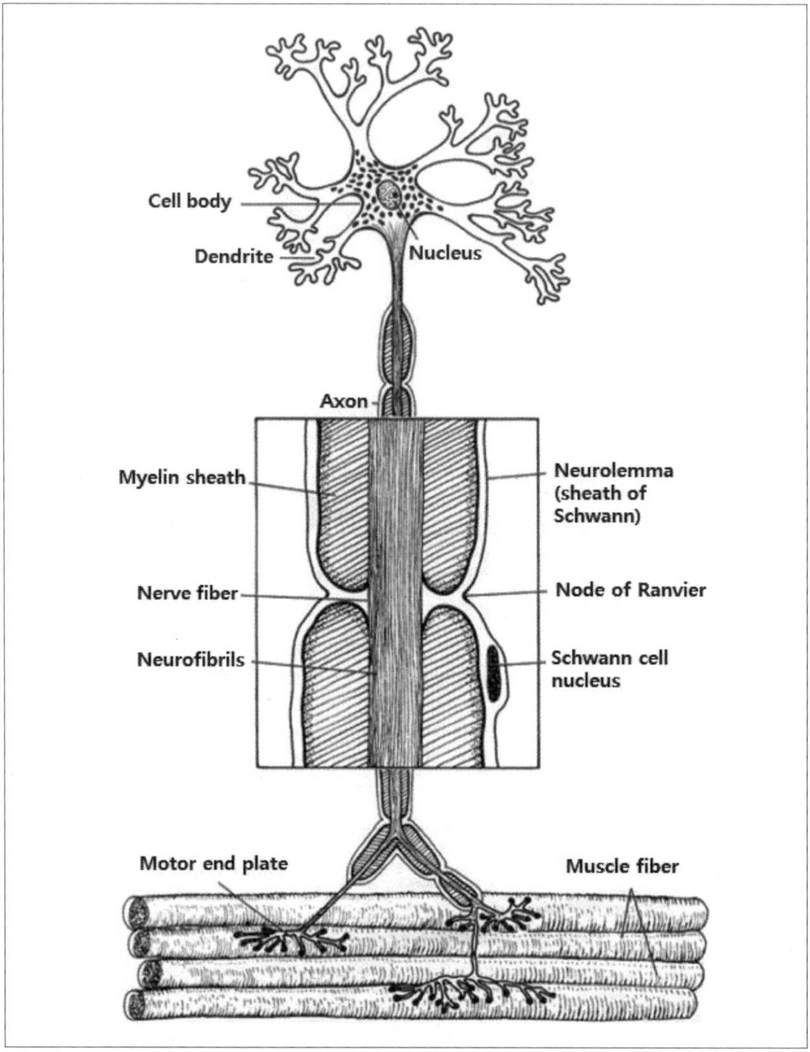

Figure 1.13. Diagrammatic illustration of a neuron and the motor end plates associated with it. The magnification of the axon (center) shows details of the nerve fiber, neurofibrils, myelin sheath, Schwann cell nucleus, and node of Ranvier.

industry is the toxin produced by Clostridium botulinum.

In the muscle, nerve trunk consisting of a group of axons can be observed as fine silvery lies because they are covered with a sheath of dense connective tissue. This arrangement helps protect the axons and provides strength to the structure. Small, peripheral nerve fibers are covered with Schwann cells that help speed up the electrical pulse going through the nerve whereas large fibers are often covered with a myelin sheath which is coming from the Schwann cells. Therefore, nerve fibers are commonly referred to as myelinated and non-myelinated fibers.

4. Epithelial Tissue

From an embryonic development standpoint, epithelial tissue is designed to serve as the interface between the body and the outside world. Hence, it consists of the skin and the lining of the digestive system. It also contains some other specialized components that will be described below.

The skin (Figure 1.14) serves as a protective layer that prevents microorganisms from entering the body and protects the body from environmental stresses such as drying. It also protects the body against mechanical damage and serves a major role in insulation and heat regulation. In general, the two major parts of the skin are the epidemis, which is the ectodermal portion, and the dermis, which is the mesodermal portion. A unique structure of poultry skin is the feathers,

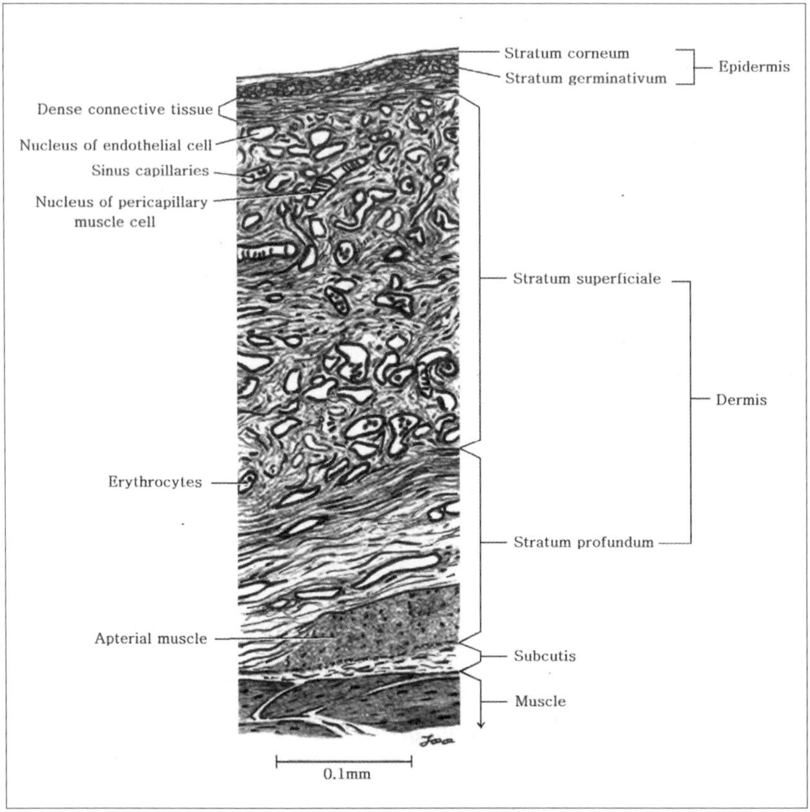

Figure 1.14. Structure of the skin section stained with hematoxylin and eosin.

which are a complex derivative of epithelial tissue. Feather size varies greatly with the longest tail feathers of a roaster being about a 1,000 times longer than the feathers on its eyelids.

Of the four tissue types present in the animal body, quantitatively, epithelial tissue contributes the least to meat. However, the characteristic flavor and crispness of fried chicken, for example, is partly due to properties of epithelial tissue and its underlying connective tissue. Epithelial tissue forms the linings that cover the surfaces of the slaughter and processing. Most of the epithelial tissue remaining in the carcass is associated with blood and lymphatic vessels, but this tissue also is present in many edible organs, such as the kidney and liver, as well as in hides and skins.

Epithelial tissue is characterized as having little extracellular material and is classified according to cell shape and the number of layers forming the epithelium. Cell shapes vary from thin, flat cells to

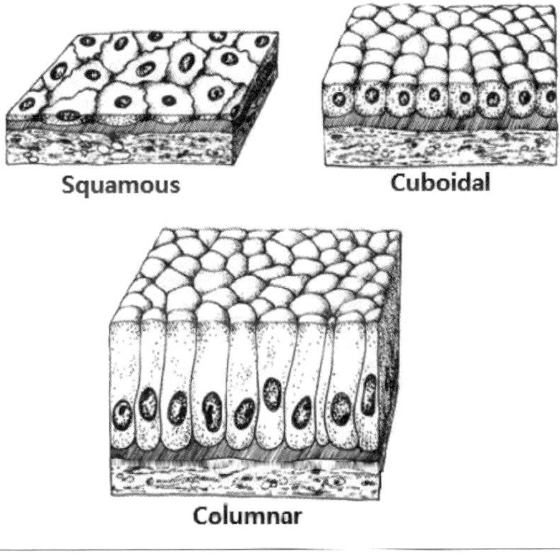

Figure 1.15. Diagrammatic illustration of squamous, cuboidal, and columnar epithelial tissue.

tall columnar cells (Figure 1.15). Some cells form single layers, while others form stratified layers. In either case, they maintain extensive intercellular contracts, thus forming cohesive cellular sheets that cover

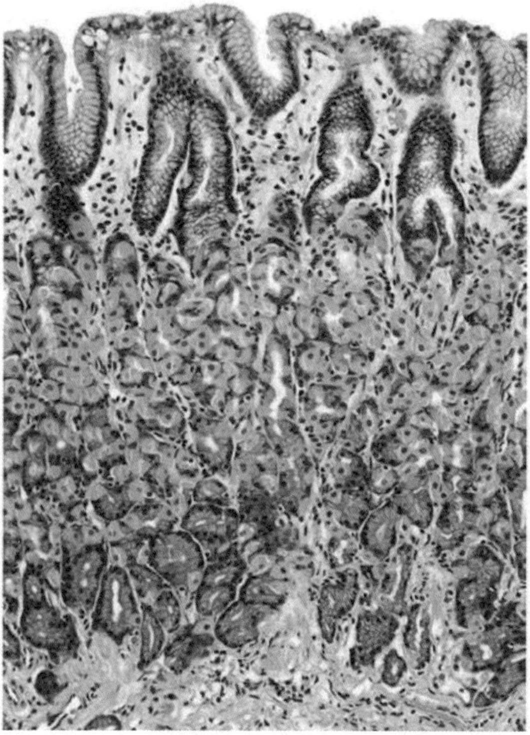

Figure 1.16. Classification of cells found in epithelial tissue is based on their shape. Showing here columnar epithelium cells lining at the top of the gut wall, as these many cells can efficiently absorb nutrients. The skin on the other hand has usually flat thin cells for protection.

surfaces and line cavities. Additionally, the surface of the epithelium may be modified to increase their efficiency in performing such functions as protection, secretion, excretion, transport, absorption, and sense perception.

Epithelial tissue is commonly characterized by cell shape and the number of cell layers (Figure 1.16). Epithelial cells are usually laid down with little extracellular material. Cell shape can vary from elongated, columnar-type cells to very thin, flat cells called squamous cells. In addition, cuboidal cells also form single or multiple layers on external or internal surfaces of the body.

Other body that contain epithelial tissue are the lining of the digestive system, liver, and kidney. In organs such as the liver and kidney, the cells secrete different enzymes than those of the digestive system, where they absorb nutrients from the gut and are usually columnar in shape to increase the number of cells in contact with food and make nutrient absorption more efficient.

Chapter 2.

Microstructure of Muscle Fiber

1. Muscle Fiber

Most animal muscle is roughly 75% water, 20% protein, and 5% fat, carbohydrates and assorted proteins. Muscles are made of bundles of cells called fibers. Each cell is crammed with filaments made two proteins which is called actin and myosin. Muscle fibers constitute 75 to 92 percent of total muscle volume. The remaining volume consists of connective tissue, blood vessels, nerve fibers, and extracellular fluid. Mammalian and avian skeletal muscle fibers are long, multinucleated,

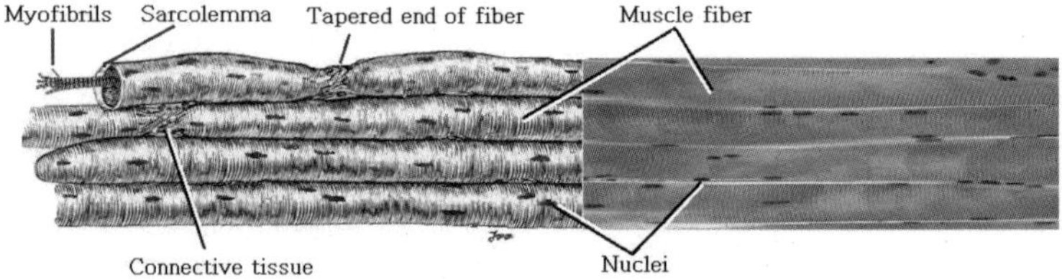

Figure 2.1. Drawing of skeletal muscle fibers showing structural features and their longitudinal orientation.

unbranched, threadlike cells that taper slightly at both ends, as shown in Figure 2.1. Although fibers may attain a length of many centimeters, they generally do not extend the length of the entire muscle. Muscle fibers vary considerably in diameter, ranging from 10 to more than 100 μm within the same species, and even within the same muscle.

2. Sarcolemma

The membrane surrounding a muscle fiber is called the sarcolemma (Figure 2.2). The prefix sarco- is derived from the Greek word sarx or sarkos, which means flesh, and the suffix-lemma, is the Greek word for husk. Sometimes called the myolemma, the sarcolemma is similar to a

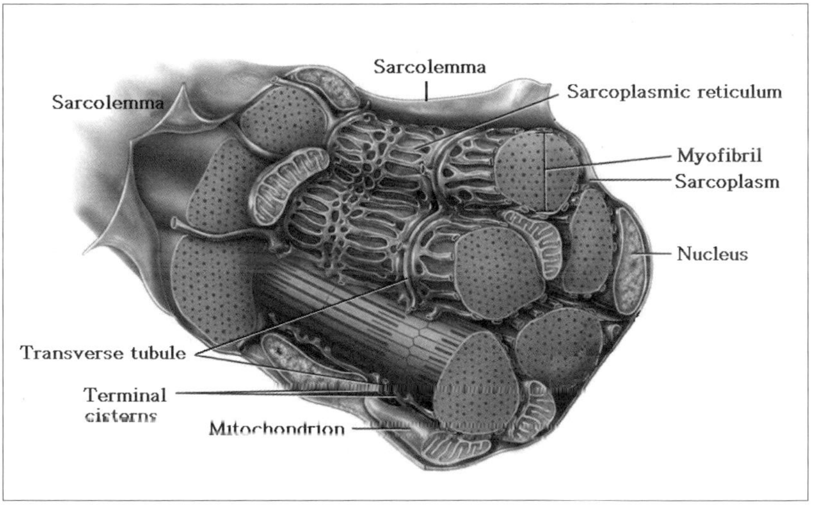

Figure 2.2. Drawing of sarcolemma surrounding a muscle fiber

typical plasma membrane but has specialized functions for the muscle cell. The sarcolemma also contains an extracellular matrix consisting of various polysaccharides which allows the cell to anchor into the tissues that build and support muscle fibers.

Typically, the sarcolemma connects the basement membrane which surrounds all connective tissues, or the other muscle cells, creating a very strong fiber which can contract together. Periodically, along the

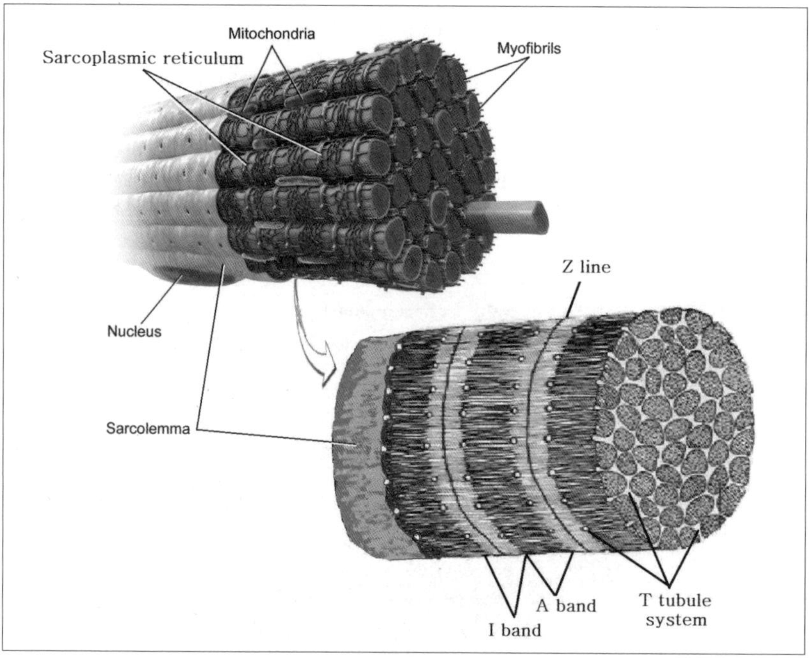

Figure 2.3. Drawing of a cross-section through a muscle fiber at the A band-I band junction, showing the transverese tubule system in mammalian skeletal. Invaginations of the sarcolemma are shown along the fiber and around it's entire circumference.

length of the fiber and around its entire circumference, invaginations of the sarcolemma form a network of tubules called transverse tubules (Figure 2.3). These tubules are sometimes referred to as the T-system or T-tubules.

3. Sarcoplasm

The cytoplasm of muscle fibers is called sarcoplasm (Figure 2.4). It is water solution containing ATP and phosphagens, as well as the enzymes and intermediate and product molecules involved in many metabolic reactions. The most abundant metal in the sarcoplasm is potassium.

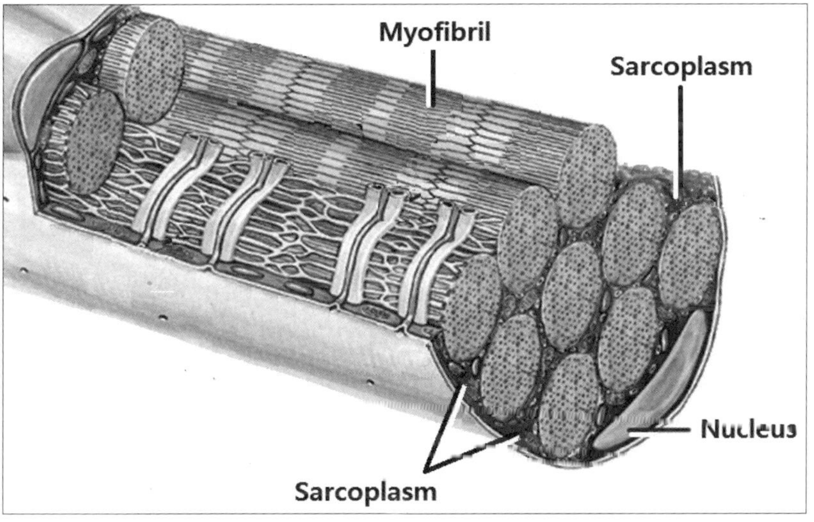

Figure 2.4. Drawing of skeletal muscle fibers showing sarcoplasm's location in microstructure.

Sodium and magnesium are present in lower concentrations. Most of the calcium of muscle is bound to proteins or stored in the sarcoplasmic reticulum. Contraction is initiated by the release of calcium ions (Ca^{2+}) upon the depolarization of the membrane, which is induced by nerve impulses. In addition, sarcoplasm of skeletal muscle contains lipid droplets, glycogen granules, ribosomes, numerous proteins, nonprotein nitrogenous compounds, and a number of inorganic constituents.

4. Nuclei

Muscle fibers are multinucleated; the nuclei lie at the surface near the sarcolemma and are best viewed as to their characteristic position in transverse section. A fiber several centimeters long may have several hundred nuclei with a regular distribution about every 5 μm along its length, except at tendinous attachments where the nuclei are more concentrated and more irregularly distributed. The concentration of nuclei also increases in the vicinity of the motor end plate. In mammalian muscle, nuclei are located at the periphery of the fiber, just beneath the sarcolemma, as shown in Figure 2.5. In contrast, the nuclei of fish skeletal muscle are usually centrally located within the fiber. Nuclei are ellipsoidal in shape, with their long axis oriented parallel to the long axis of the fiber. These features of muscle fiber nuclei can best be visualized by examining the transverse and longitudinal position of nuclei illustrated in Figure 2.1 and 2.5.

The number of muscle nuclei increases more than 20-fold from birth to adulthood. There is a high correlation between the increase in nuclei number and the increase in muscle weight. It is possible that the addition of a unit of muscle DNA will result in the addition of a unit of muscle protein in a normal animal. Nuclei that are in a muscle cell do not divide. The increase in muscle nuclei number during muscle growth results from fusion of satellite cells with the muscle cell. Satellite cells are mononucleated cells located between the muscle cell and its surrounding basal lamina. Satellite cells can divide, and some of the resulting cells will fuse with the muscle cells. This results in an

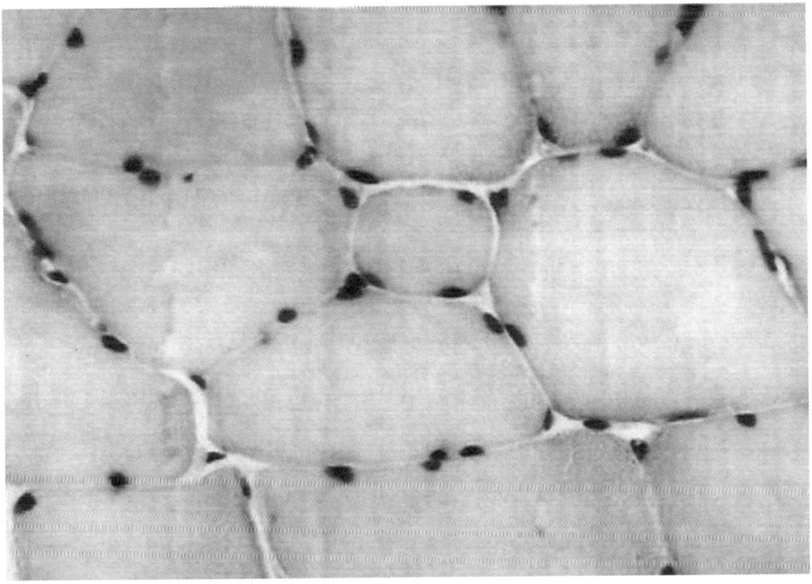

Figure 2.5. Photomicrograph of a cross-section of skeletal muscle fibers of the pig, showing their polygonal shape and the peripheral position of nuclei (×630).

increased number of nuclei in the muscle cell. The cessation of muscle growth or a reduced growth rate has been correlated with a reduction in nuclear proliferation. The number of satellite cells per muscle cell is reduced during aging.

5. Myofibrils

Myofibrils are the elongated intracellular contractile elements which measure about 1 μm in thickness and are directly responsible for the characteristic banded or striated pattern of skeletal muscle. In most mammalian skeletal muscle, the long axis of a myofibril is parallel to the long axis of the fiber. Myofibrils are bathed in sarcoplasm and extend the entire length of the muscle fiber. Diagrams and electron photomicrographs of the myofibril are shown in Figure 2.4 and in Figure 2.7 (D and E), respectively. Muscle fibers of meat animals with diameters of 50 μm have at least 1000 and could have as many as 2000 (or more) myofibrils.

This striation arises from the orderly alignment of the anisotropic and isotropic segments of the myofibrils and resembles the stacking of light and dark segments. With polarized light, the discs that seem darker with the ordinary light microscope are anisotropic (birefringent) while those that appear lighter with the light microscope are isotropic. As a result of these properties, the darker discs are called A bands, and the lighter discs are called I bands. A dark line termed the Z line bisects

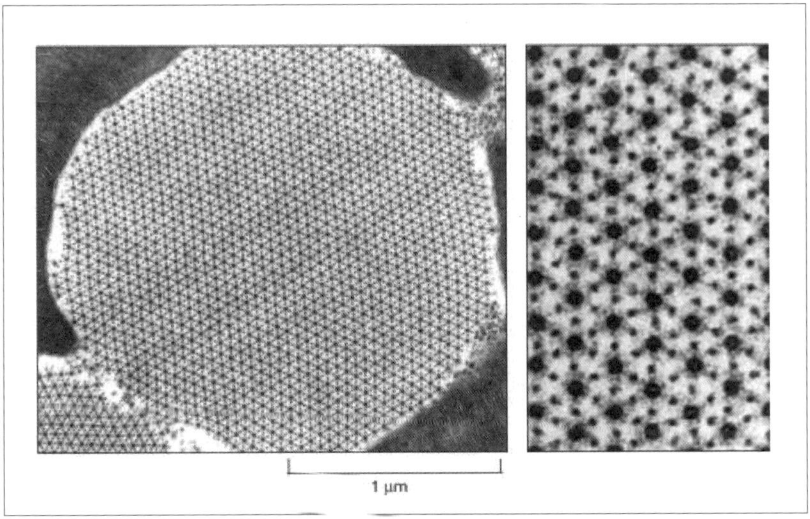

Figure 2.6. Electron photo-micrograph of a cross-section of a myofibril, showing the thick and thin filaments and their hexagonal arrangement.

the I band. A light area in the center of the A band is known as the H band, and bisecting this band is a dark M line. The sarcomere or unit of muscle structure is the area between two adjacent Z line (Figure 2.7D; 2.7E and Figure 2.9).

Photo-micrographs of myofibrillar cross-sections reveal a highly organized array of dots that have two types sizes (Figure 2.6). These dots represent the two types of myofilaments within myofibrils (shown in longitudinal section in Figure 2.7D and E).

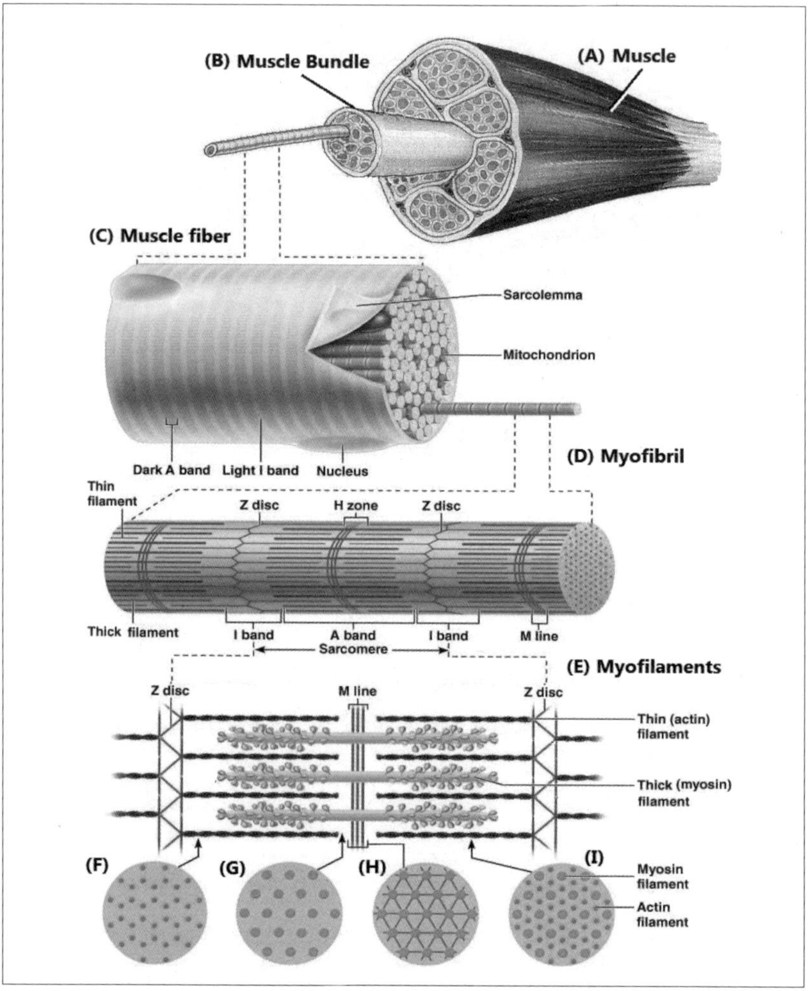

Figure 2.7. Diagram of the organization of skeletal muscle the gross structure. (A) skeletal muscle, (B) a bundle of muscle fibers, (C) a muscle fiber, showing the myofibrils, (D) a myofibril, showing the sarcomere and it's various bands and lines, (E) a sarcomere, showing the position of the myofilaments in the myofibril, (F-I) cross-sections showing the arrangement of the myofilaments at various locations in the sarcomere.

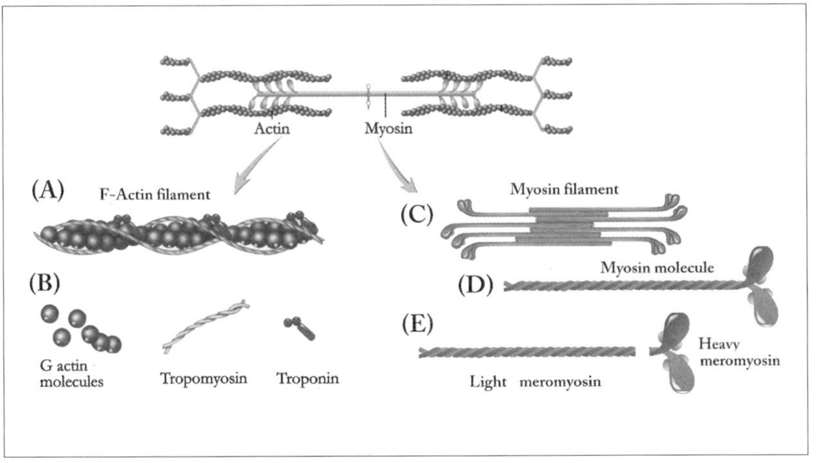

Figure 2.8. Diagram of the organization of myofibrils gross structure to the molecular level. (A) an actin filament, composed of two-F-actin chains coiled about each other, (B) G-actin molecules, (C) a myosin filament, showing the relationship of the heads to the filament, (D) a myosin filament showing the head and tail regions and, (E) the light meromyosin (LMM) and heavy meromyosin (HMM) portions of the myosin molecule.

5.1. Myofilaments

Myofilament is the term for the chains of (primarily) actin and myosin that pack a muscle fiber and these actin and myosin are located in a thick and thin filaments of the myofibril. In longitudinal section, thick filaments are aligned parallel to each other and arranged in exact alignment across the entire myofibril. Likewise, thin filaments are aligned precisely across the myofibril, parallel to each other and to the thick filaments. This arrangement of myofilaments, and the fact that thick and thin filaments overlap in certain regions along their longitudinal axes, account for the characteristic banding or striated appearance of the myofibril (Figure 2.8 and 2.9).

The transverse striations of fibrils are caused by the regular longitudinal arrangement of sets of thick filaments (10 to 12 nm in diameter) and thin filaments (5 to 7 nm in diameter). In a transverse section cut through overlapping thick and thin filaments, each thick filaments is surrounded by six thin filaments, although this hexagonal lattice may change to a tetragonal lattice when sarcomeres are stretched (Figure 2.7 and 2.9).

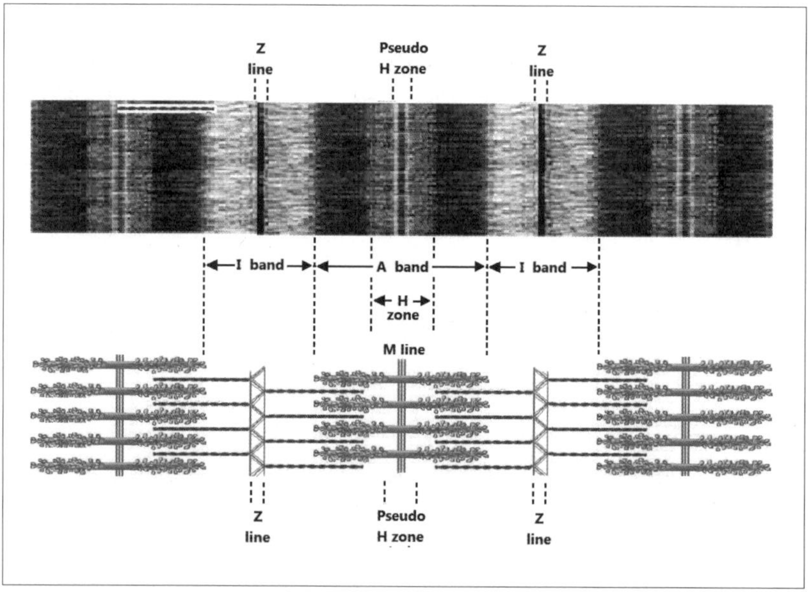

Figure 2.9. A drawing adapted from an electron photomicrograph, showing portions of two myofibrils and a sarcomere and a diagram corresponding to the sarcomere, identifying it's various bands, zones, disks, and lines.

The transverse striations on muscle fibers are due to the precise alignment of A and I bands on fibrils within the fiber. In most stained preparations for light and electron microscopes, A bands appear darker than I bands (the reverse appearance to that seen with polarized light). This is because the birefringent A bands contain a greater density of protein. A bands also appear darker than I bands when unstained preparations are observed with a phase contrast light microscope (a special type of microscope that shadows differences in glassy properties of the specimen). In the electron micrograph (Figure 2.10), the fibrils are running from the top left to the bottom right, so that the transverse striations are from the top right to the bottom left.

Figure 2.10. Striations that appear bright in polarized light are anisotropic or A bands while those that appear dark are isotropic or I bands

5.2. Sarcomere

The unit of the myofibril spanning two adjacent Z disks is referred to as a sarcomere. The sarcomere includes both an A band and the two half I bands located on either side of the A band. Sarcomeres are the smallest contractile units and serve as the basic force-producing machinery of striated muscles. Sarcomeres are renewed continuously during muscle growth and physiological turnover and are assembled from an immense number of protein subunits into a very large and most regular macromolecular complex. The structure of the sarcomere is very interconnected and at the same time extremely ordered. The sarcomere is the repeating structural unit of the myofibril, and it is also the basic unit where muscle contraction and relaxation occur.

Other features of the fibril are detectable by light microscopy under optimum conditions, but these details are seen more clearly by electron microscopy. A thin Z line or disc occurs at the middle of the I band. The transmission electronmigrograph above is not the only way to look at transverse striation. The image below (Figure 2.11) is a scanning electron micrograph, and the fibrils run horizontally across the image, showing transverse striation vertically. This image does not look the same as in Figure 2.10. In the scanning micrograph can see the sarcomeres because of the raised region of the Z-line attached to the cytoskeleton (desmin).

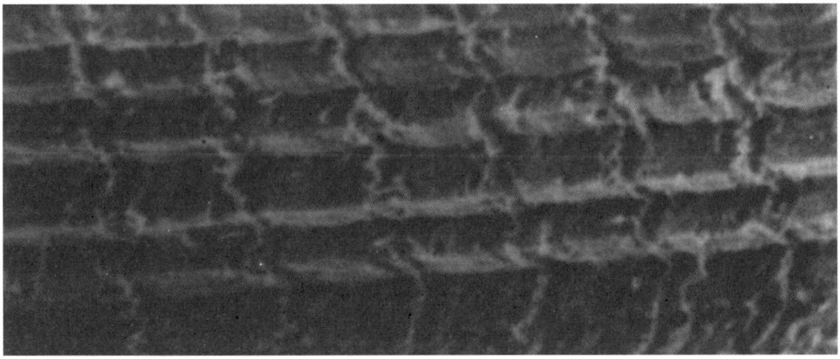

Figure 2.11. The repeating unit of a regular series of transverse striations of the sarcomere and the structural unit from Z-line to the other Z-line.

5.3. Myosin and Actin Filament

The main constituent of the thick filaments is myosin (Figure 2.8). Each thick filament is composed of about 250 molecules of myosin. Myosin has two important roles: a structural one, as the building block for the thick filaments, and a functional one, as the catalyst of the breakdown of ATP during contraction and in its interaction with actin as part of the force generator of muscle. Thick filaments of vertebrate muscles are approximately 14 to 16 nm in diameter and 1.5 μm in length. Since the predominant protein in thick filaments is myosin, they are referred to as myosin filaments. Myosin filaments are held in position by other protein, some of which are located in the M line (Figure 2.14).

Actin, which constitutes about 25 percent of the protein of myofilaments, is the major component of the thin filaments in muscle (Figure 2.8). Thin filaments are about 6 to 8 nm in diameter and extend approximately 1.0 μm on either side of the Z disk. Because actin is the most abundant protein in the thin filament, they are referred to as the actin filaments. Actin filament components associate with nascent Z-disk structures when myofibrils are formed. In thin filaments also contain other proteins of tropomyosin and troponin. Tropomyosin has a molecular weight of 37 kDa and is constructed of two α-helical chains arranged as a coiled-coil rod. The primary function of tropomyosin is together with the troponins to regulate the interaction of myosin and actin filaments during the force generation. The troponins are a complex of three proteins: troponin-C (18 kDa), troponin-I (20 kDa) and troponin-T (30 kDa) and are present at regular intervals along the entire length of the actin filament.

5.4. Z Disk Ultrastructure

Z-disks are the primary conduits of the force generated by contraction and constitute an anchoring point for the actin filaments as well as for titin and nebulin. The Z-disks of adjacent myofibrils are aligned which facilitates coordination of contractions between individual myofibrils. The Z-disk is a complex protein network. Each actin filament connects to four Z filaments that pass obliquely through the Z disk. Each of the four Z filaments then connects with an actin filament in the adjacent sarcomere, as shown diagrammatically in Figure 2.12.

α-Actinin (97 kDa) is a major component of the Z-disk, function as an actin cross-linking protein and helps to stabilize the sarcomere

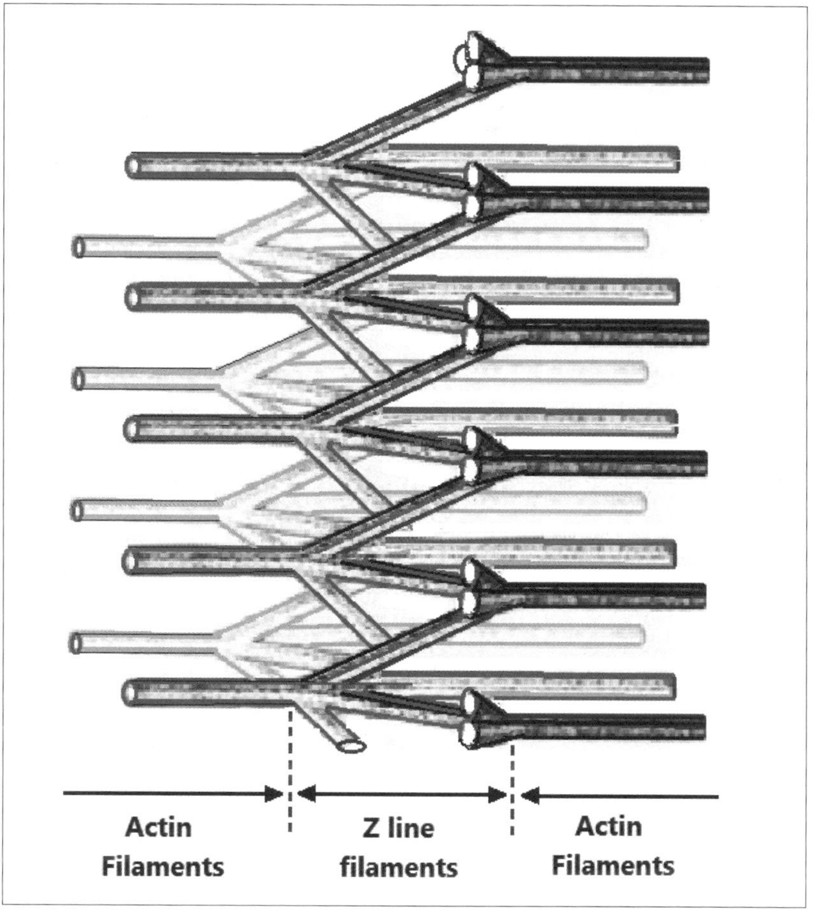

Figure 2.12. Diagrammatic representation of Z filaments and their attachment to actin filaments.

structure during contraction. Other Z-disk associated proteins include muscle LIM protein (23 kDa), filamin (around 300 kDa), telethonin (19 kDa) and MURF-3. Intermediate filaments and their associated proteins act as linkages to the cytoskeletal network and proteins such as desmin (52 kDa) are present at the periphery of the Z-disk and form the link between adjacent myofibrils. Thus the Z-disk is strongly involved in stabilizing the sarcomere against lateral shear forces as well as transmitting axial forces along the muscle fiber.

6. Protein of the Myofibril

The major proteins of the myofibril are myosin, actin, tropomyosin, troponin, the actinins, C-protein, M-protein, F-protein, I-protein, myomesin, and the filament protein desmin, Z- protein, nebulin, and titin. Actin and myosin constitute the major contractile proteins while tropomyosin and troponin are the major regulatory proteins which is to regulate actin-myosin interaction during muscle contraction. Titin and nebulin which is known as cytoskeletal protein, provide the alignment of myofilaments during myofibril and sarcomere formation. In the following discussion of the individual proteins, general properties based on findings in various species will be presented.

6.1. Major Contractile Proteins

Actin is a globular protein made up of a single polypeptide chain which binds one molecule each of nucleotide (ATP or ADP) and divalent cation (Ca^{2+} or Mg^{2+}) per mole of monomer. Actin, which is constitutes about approximately 20 percent of the myofibrillar proteins is the major component of the thin filaments in muscle. An individual molecule of actin is a single protein chain coiled to form a globular (spherical) shaped molecule approximately 5.5 nm in diameter. Actin in this form, called globular actin or G-actin which has one calcium or magnesium ion and one molecule of ATP bound to it (Figure 2.8B). The fibrous nature of the actin filament occurs when G-actin monomers polymerize to form F-actin (fibrous actin), as shown in Figure 13B and Figure 2.8A.

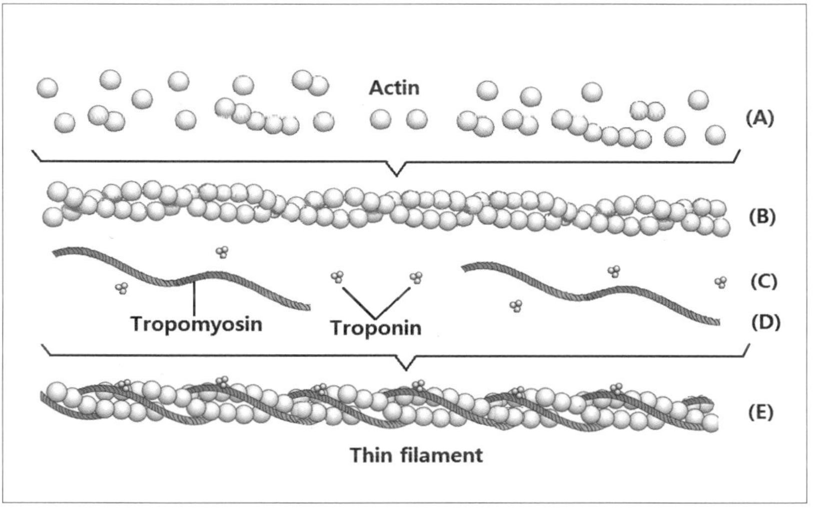

Figure 2.13. Diagrammatic representation of the construction and fine structure of an actin filament, showing the relation of tropomyosin and troponin to its (A) G-actin molecules, (B) two chains of polymerized actin (F-actin) coiled about each other, (C) pellet-shaped troponin molecules form two long strands of tropomyosin, one strand being associated with each of the F-actin chains, (D) long thin molecules of tropomyosin. The periodic location of troponin along the actin filament every seven G-actin molecules, and its association with the tropomyosin strand, are also illustrated (E).

When the G-to-F transformation takes place, the ATP bound to G-actin breaks down, releasing inorganic phosphate (Pi) and leaving an adenosine diphosphate (ADP) molecule bound to each actin unit. In F-actin, the G-actin monomers are linked together in strands of F-actin are spirally coiled around one another. They give rise to a helical structure that can be viewed as a double or single helix and this is the characteristic of the actin filament. Actin is believed to be directly involved in the process of contraction because the cross bridges can become attached to it.

Myosin is fibrous protein that constitutes approximately 45 percent of myofibrillar protein. The individual myosin molecule contains two major protein chains and four small ones, the entire molecule being about 160 nm in length and asymmetrically shaped. The rodlike tail region, about 120 nm long, consists of two chains of protein, which is known as an α-helix, together forming a coiled-coil structure. The thickened end of the myosin molecule is usually referred to as the head region, and the long, thin portion that forms the backbone of the thick filaments is called the rod or tail portion. The portion of the molecule between the head and the rod regions is called the neck. The head region of the molecule is double headed, and it projects laterally from the long axis of the filament (Figures 2.14B and 2.8D). As mentioned in previous part, myosin has two important roles one of them as the building block for the thick filaments while the other function as the catalyst of the breakdown of ATP during contraction. At the other end of the molecule, the two protein chains form two globular headlike regions that have the ability to combine with the protein actin and carry the enzymatic sites for ATP hydrolysis.

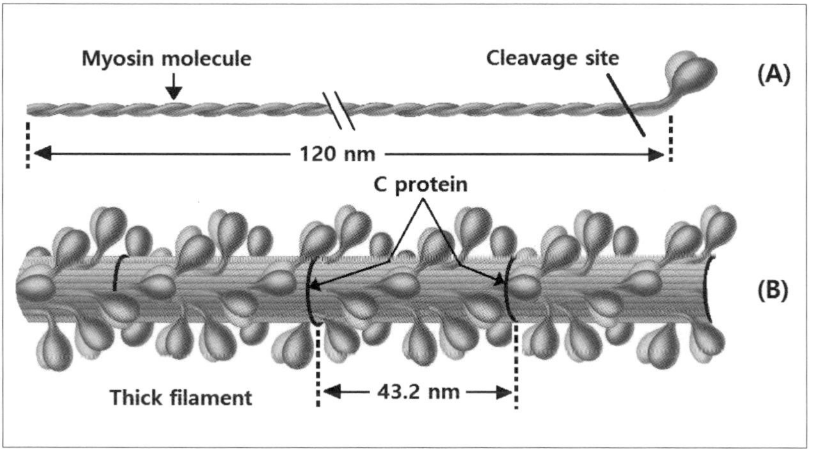

Figure 2.14. Diagrammatic representation of the construction and fine structure of the myosin filament. (A) The myosin molecule showing the head and tail, and the cleavage site. (B) A myosin filament, showing the projecting double heads and periodic (every 43.2 nm) encircling band of C protein.

6.2. Major Regulatory Proteins

Tropomyosin constitutes approximately 5 percent of myofibrillar protein and has a molecular weight of 37 kDa. Tropomyosin is constructed of two α-helical chains arranged as a coiled-coil rod and lies in close contact with the actin filament. Each tropomyosin molecule is in contact with seven G-actin molecules in the actin filament. The relationship of tropomyosin to the actin filament is illustrated in Figure 2.13D.

Troponin also constitutes approximately 5 percent of myofibrillar protein. It is a complex of three different protein subunits. One troponin complex is bound to every tropomyosin molecule. A troponin molecule is located approximately every 40 nm along the actin filament, as shown in Figure 2.13D. There is one molecule of troponin for every seven or eight G-actin molecules along the actin filament.

Troponin and tropomyosin are both involved in the regulation of the contraction and relaxation of muscles. One of the subunits (TnC) is the receptor for Ca^{2+} released from the sarcoplasmic reticulum on activation of the muscle. It is thought that calcium binding then causes further structural changes in the interaction of actin, tropomyosin, and another troponin subunit (TnI) that lead to contraction by activating the actin-myosin interaction.

6.3. Cytoskeletal Proteins

Titin and nebulin, the two giants of the sarcomere, are muscle specific proteins that play key roles in organizing sarcomeres (Figure 2.15). Titin is the most abundant cytoskeletal protein in muscle which comprising about 10 percent myofibrillar proteins and many of the properties of the sarcomere involves the giant protein titin (also known as connectin), and this protein is the third most abundant after myosin and actin. It extends longitudinally in each half sarcomere from the M line to the Z disk as a third filament of the myofibril. Because titin has an elastic component in the I-band region while the portion of titin in the A band is inelastic. Six titin molecules are bound to each half of the myosin filament and the protein (C protein) that encircles and stablizes the thick filament. Titin is believed to define the resting sarcomere length and to keep the myosin filaments in the center of the sarcomere. Another main function of titin is to provide elasticity (molecular spring) and can induce to conformational changes in the molecule during both active muscle contraction and passive extension.

Nebulin composes approximately 4 percent of myofibrillar proteins and an actin-binding protein of 500 to 800 kDa. It is incorporated in the actin filaments and has a length of around 1.0 μm. Nebulin extends

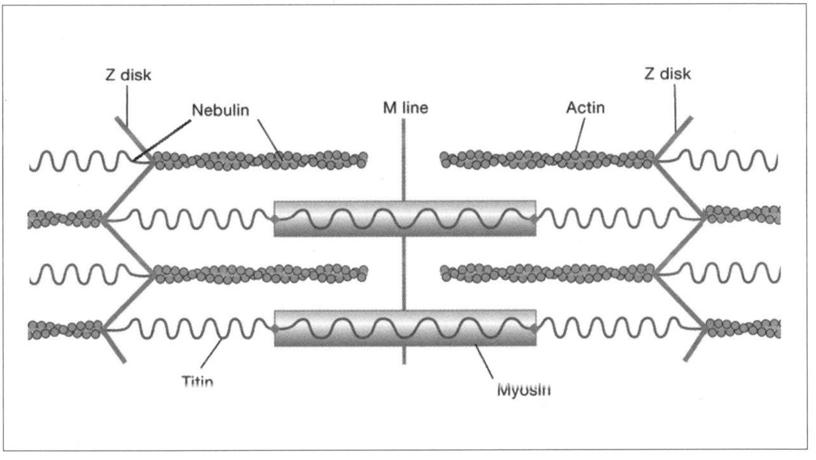

Figure 2.15. Diagrammatic representation of titin and nebulin, and their attachment to actin and myosin filaments. Titin provides elasticity and stabilizes myosin. Nebulin helps align actin.

longitudinally along the entire length of the thin filament from the A band to the Z disk. The primary structure of nebulin is composed of 185 α-helical repeats and the N-terminus extends to the ends of the actin filaments, whereas the C-terminus is partially inserted into the Z-disks. In developing muscle, it is believed to play a role to organize the structure of the sarcomere. In mature muscle, it serves as a template for assembly and/or a scaffold for stability of thin filaments. Nebulin can associate directly with titin at the Z-disk and these two proteins may act together to determine the width of the Z-disk.

Obscurin is the most recently discovered of the giant proteins in the sarcomere. This protein has been difficult to characterize due to its large size and insolubility. With the molecular weight of 720 to 900 kDa, this protein is not present within the sarcomere but surrounds them at the level of the Z-disk and M-line. Obscurin interacts with myosin and titin and also with proteins in the sarcoplasmatic reticulum membranes.

C protein constitutes approximately 2 percent of myofibrillar proteins. It has a molecular weight of 135 kDa and a relatively high proline content (71/1000 residues). At lower ionic strengths, C protein copolymerizes with myosin filaments but does not appear to be bound to the myosin rod along the length of the thick filament with a specific periodicity. H protein, approximately 1 percent of myofibrillar protein, also is present in this region. M protein with a molecular weight of 165 kDa is found at the M-line of the thick filaments together with Myomesin and skelemin. This proteins constitute less than 2 percent in the myofibrillar proteins. F protein has a molecular weight of 121 kDa and its amino acid composition is very different from all of the myosin associated proteins. This proteins bind to myosin filament but the presence of C protein can inhibit the binding ability. While I protein high in aspartic and glutamic acid with the molecular weight of 50 kDa. It is located at the A band and in the absence of calcium ion, I protein inhibits actomyosin ATPase activity. Creatine kinase also is present at the M line but does not have a role in this structure. Alpha-actinin and Cap Z, which represent 2 percent and less than 1 percent, respectively, of myofibrillar proteins, are integral components of the Z disk.

Other proteins appear to be part of the Z-line are desmin, Z protein, vimentin, paranemin, and synemin. This proteins are components of thin or intermediate filaments peripheral to the disk itself. Desmin is the same as skeletin which has a molecular weight of 55 kDa. Two types of isoforms, α and β, with isoelectric points of 5.65 and 5.70, respectively, and soluble at low pH. Z protein has the same molecular weight as desmin, however, thus far, no interaction between Z protein and actin or any other proteins of the Z-disk has been observed. While, vimentin, paranemin and synemin are only a few of the many intermediate filament-like proteins that have been identified in muscle fibers.

7. Sarcoplasmic Reticulum and T Tubules

The other membrane system that surrounds each myofibril is the sarcoplasmic reticulum which forms a cufflike structure surrounding a myofibril (Figure 2.16 and Figure 2.2). The portion in contact with the transverse tubule (T tubules) forms an enlarged sac called the terminal cisternae. The sarcoplasmic reticulum and transverse tubules, although usually discussed together, but these two components are two separate and distinct membrane structures. The T tubules are associated with the sarcolemma, whereas the sarcoplasmic reticulum is an intracellular membrane structure.

The sarcoplasmic reticulum controls the level of calcium ions in the sarcoplasm. In muscle, calcium ion is the trigger which causes the myofibrils to contract and develop force. The sarcoplasmic recticulum, an intricate and interconnected intracellular membrane system, is believed to maintain the cytoplasmic calcium concentration at 2 x 10-7 M during rest and release calcium (up to 10-5 M) in response to a signal from the T-tubules. The calcium diffuses to the myofibrils, binds to troponin on the thin filament and thereby allows myosin and actin to interact. The sarcoplasmic recticulum then rebinds the calcium until a new contraction cycle is initiated.

In most vertebrates each transverse tubule has two cisternae closely associated with it, forming a three-element complex called a triad. The number of triads per sarcomere depends on the species; for example in

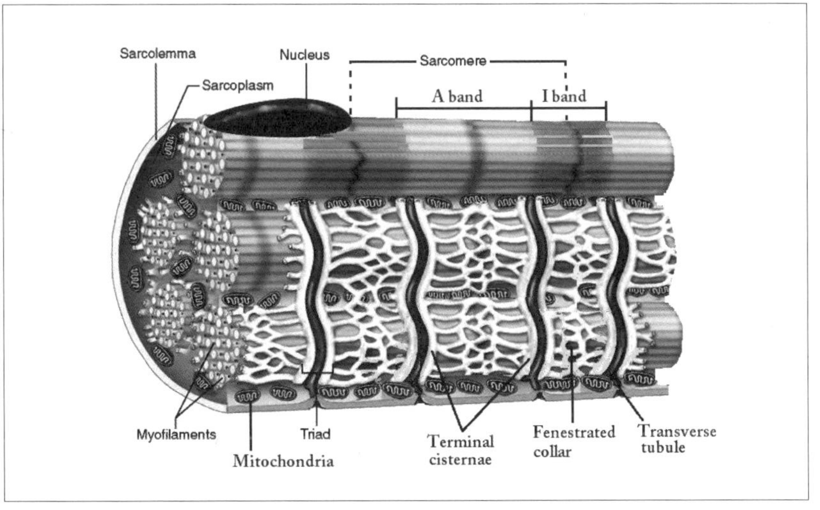

Figure 2.16. Ultrastructure of a group of myofibrils, showing the sarcoplasmic reticulum and transverse tubules and their relation to the myofibrils of mammalian skeletal muscle.

mammalian muscle there are two triads. In fishes and crustacean only one cisternae is associated with each transverse tubule, thus forming a dyad. The terminal cisternae apparently are the sites from which the calcium ions are released when the muscle is stimulated, and the longitudinal tubules are the sites at which calcium ions are effectively removed from the sarcoplasm. The removal of calcium ions (Ca^{2+}) from the sarcoplasm is accomplished by a protein that catalyzes the breakdown of ATP, making the free energy of hydrolysis available for the energy-requiring process of Ca^{2+} transport.

8. Mitochondria

Mitochondria are oblong organelles located in the sarcoplasm and often referred as the powerhouse of the cell because it is the place where the majority of cellular ATP is produced through a process called oxidative phosphorylation. The distribution of mitochondria in muscle was variable, but usually, they were found longitudinally oriented in the spaces between myofibrils, immediately under the sarcolemma or clustered at the poles of the nuclei (Figure 2.2 and 2.16). Quite often they were found in a horizontal plane, at the level of the Z-line or the A-I junction. The cessation of oxygen delivery to the muscle following exsanguination impedes the mitochondria's ability to produce ATP through oxidative phosphorylation. Therefore, mitochondria are often considered irrelevant to the process of converting muscle to meat. However, mitochondria do not stop its process immediately postmortem, rather they maintain functionality and structural integrity for several hours. Mitochondria may influence postmortem metabolism by altering the biochemical and the energetic properties of the muscle. In addition, mitochondria are involved in cellular calcium ion homeostasis, apoptosis, and meat color stability.

Mitochondria consist of two membranes (inner and outer), hence creating two separate mitochondrial compartments: the inner membrane space between the outer and the inner membrane and the matrix, which is bound by the highly convoluted inner membrane. The outer membrane is highly porous and permeable to most ions and small molecules. The outer membrane, which was also isolated as a relatively pure fraction, contained the ancillary mitochondrial enzyme systems that implement citric acid cycle oxidation, fatty acid oxidation, fatty acid elongation, and substrate level phosphorylation. In contrast, the inner membrane is impermeable to the passage of nearly all ions and molecules, unless mediated by a membrane transport protein. The inner membrane, which could be isolated in a relatively pure form, contained the four complexes of the electron transfer chain as well as enzymes such as ATPase.

9. Lysosomes

Lysosomes are small vesicles located in the sarcoplasm that contain a number enzymes collectively capable of digesting the cell and its contents. Included among these lysosomal enzymes are a group of proteolytic enzymes known as cathepsins. The lysosomal cathepsins are present in all mammalian cell types, with the exception of enucleated red blood cells. The concentration of lysosomes and hence the proteases, varies in different cells and tissues and is particularly high in liver, spleen, kidney and macrophages. In general, the lysosomal enzymes are small protein (20~40 kDa), optimally active at acidic pH values, and unstable at neutral or alkaline pH values. Of the 13 lysosomal proteases identified thus far, seven have been shown to exist in skeletal muscle cells. These include cathepsin A, B1, C, D, H, L, and lysosomal carboxypeptidase B. From these, the best characterized in degradation of myofibrils are cathepsin B, D, H, and L.

10. Golgi Apparatus

The Golgi apparatus (or knows as Golgi complex) is one of the most elaborate organelles in the cell. It consists of flattened membrane sacs called cisternae, which are usually but not always organized into polarized stacks. The Golgi apparatus contains multiple classes of cisternae that differ in structure, composition, and function, but there is no consensus about the number and definition of theses classes.

Golgi cisternae can be divided into three classes that correspond to functional stages of maturation (Figure 2.17). First, cisternae at the cisternal assembly stage receive COPII vesicles from the ER and recycle components to the ER in COPI vesicles. At this stage, new cisternae are generated. Second, cisternae at the carbohydrate synthesis stage exchange material with one another via COPI vesicles. At this stage, most of the glycosylation and polysaccharide synthesis reactions occur. Third, cisternae at the carrier formation stage produce clathrin-coated vesicles and exchange material with endosomes. At this stage, biosynthetic cargo proteins are packaged into various transport carriers, and the cisternae ultimately disassemble. Discrete transitions occur as a cisternae matures from one stage to the next. Within each stage, the

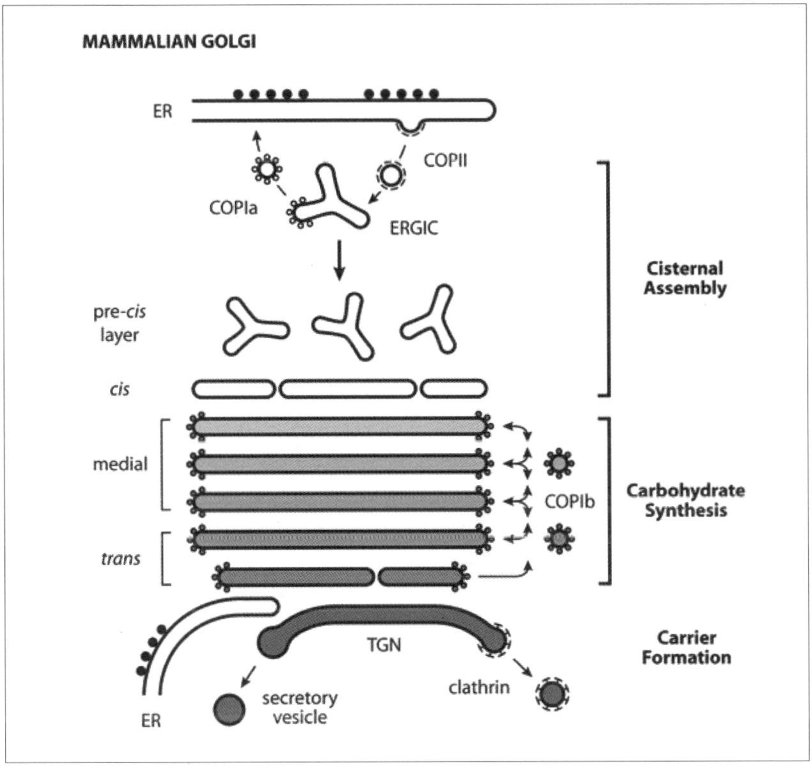

Figure 2.17. Three-stage models of the mammalian based on electron tomography of high-pressure frozen cells.

structure and composition of a cisternae can evolve, but the trafficking pathways remain unchanged. This model offers a unified framework for understanding the properties of the Golgi in diverse organisms.

Chapter 3.

Growth and Development of Carcass Tissues

A sufficient understanding of the growth and development of livestock is essential to mastering livestock evaluation because production of meat is dependent on the growth process. Growth is one of the factors that is important in the market fundamentals of all species. Even though this growth process had been widely study to investigate and improve their eating and economical value, but not all aspects of growth are fully understood, particularly the cellular mechanisms from the initiation of growth until termination of growth at maturity, as well as intrinsic and extrinsic factors of animals itself.

1. Concepts of Growth And Development

Obviously, not all tissues develop at the same rate, some are highly developed during the gestation (head), whereas other body parts are highly developed during the postnatal stage of life (muscles). Body structure and function as well as its proportions and composition change as the individual animal grows. Also, closely linked to the physical growth is the development of new structures and changes in functional capabilities. Even different species these parts seem to differ greatly, for instance, a newborn calf or lamb, their legs often do not seem to match the rest of their body, and consequently their legs do not grow as much as some of their other parts. On the other hand, baby pigs have little legs that more uniformly match their body size at birth, and gigantic ears in relation to their body. The differences in which each of these body parts grow is termed development.

The process of growth and development of different animals are complex process and the way animals develop makes it difficult to define growth. Growth is actually are a combination of height and length, or weight, and can also be described as an increase in structural tissues; bone, muscle, and the connective tissues of muscle. In addition, fat is the third tissue that is discussed with growth and development. Unlike the structural tissues its development can be the result of both hypertrophy and hyperplasia. What is the hypertrophy and hyperplasia? In Greek, hypertrophy means, increase in cell size, while hyperplasia refers to an increase in the number of cellular tissue. Consequently, the increase in size is accomplished by the following processes:

- Hypertrophy, which is enlargement of existing cells.
- Hyperplasia, which is multiplication or production of new cells.
- Accretionary growth, which is an increase in extracellular material.

It is important to distinguish between true growth, which involves an increase in structural tissues such as muscle, bone, and vital organs, and fattening, which is essentially an increase in adipose tissue. Both phenomena fit within definition of growth but must be considered separately when applied to practical animal production.

Development, which must be considered and defined along with growth, is a gradual progression from a lower to a higher stage of complexity, as well as a gradual expansion in size. Differentiation is a component of development and is the process by which cells and organs acquire completely individual characteristics. This occurs in progressive diversification of cells of the embryo into muscle cells, brain cells, liver cells, and so forth. Differentiation results in heterogeneity of form or chemical composition. Morphogenesis is the organization of various dividing cells into specific organs, each having different characteristics. Maturation is the process of becoming fully developed and reaching the highest stage of complexity. It should be recognized that different tissues and body parts grow, develop, and mature at different rates. Therefore it is difficult to pinpoint exactly when an animal becomes mature, because some tissues are still developing long after others have attained a high degree of complexity. Senescence, on the other hand, begins in individual tissues and organs when they are no longer maintained in their mature form and undergo degenerative changes without complete replacement or repair.

Growth begins with the fertilized ovum or zygote and proceeds through several distinct phases. Prenatal growth (conception to birth) may be divided into three phases: ovum, embryonic and prenatal. The prenatal period (gestation period) is approximately 113, 147 and 280 days for the pig, sheep and, cow, respectively. The ovum phase lasts from the moment of fertilization until gestation (when the three primary cell layers, ectoderm, mesoderm, and endoderm, are developed). This period may last 14 days in some meat animals.

During the embryonic phase, tissues may extend to 25 to 45 days of gestation depending on the species of meat animal. The fetal phase lasts from the embryonic phase until birth, and is characterized by different growth rates of various organs and fetus. Organs and systems such as the central nervous system, heart, liver, and kidneys, which have functional importance during fetal growth, undergo a major proportion of their growth in early stages. The digestive tract, which is not vitally important until birth, undergoes a greater proportion of its growth in later stages.

Phases of postnatal growth and development are not as distinct as prenatal phases. As an animal grows and develops from birth to maturity, continuous changes occur in its body conformation and composition. A fundamental principle of postnatal growth is that the shape of the growth curve is similar in all species and is represented by a sigmoid curve, such as that shown in Figure 3.1.

For a period of time after birth, an animal experiences a very slow growth phase. This is followed by a rapid growth phase, during which the rate of increase in size may be nearly constant and the slope of the curve remains almost unchanged. During later stage of this phase, growth rates of muscles, bones, and vital organs slow, and fattening

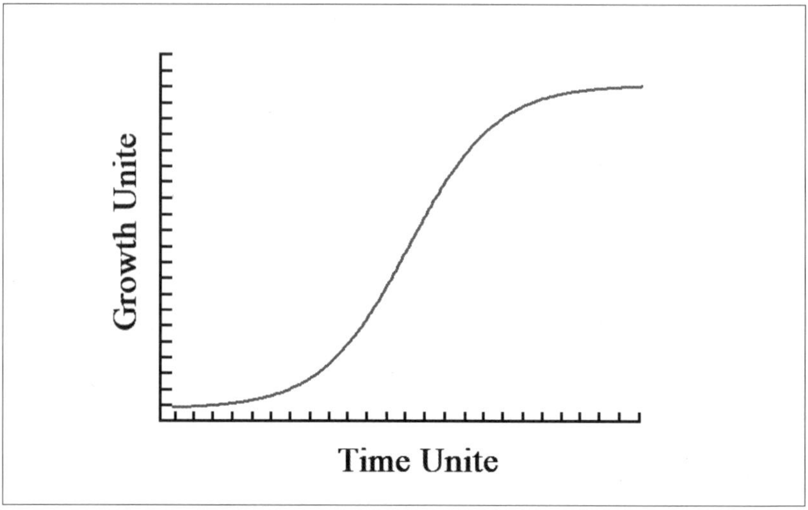

Figure 3.1. A typical growth curve.

accelerates. Finally, as the animal reaches mature size, a retardation of growth occurs.

Postnatal growth of all animal, including such extreme examples as a mouse and an elephant, fits a sigmoid curve, if appropriate time and growth units are used. Between and within species, animals having large mature size require longer time for each growth phase than do animals with relatively small mature size.

All prenatal or postnatal growth, development, and differentiation depend on changes in protein synthesis. Proteins form the structural and functional bases of every cell. tissue, and organ. Enzymes that catalyze the synthesis of body constituents are proteins. Proteins serve as means of transport and communication within cells, between cells, and between tissues. Instructions for synthesizing numerous proteins and information necessary to control protein synthesis are contained in the deoxyribonucleic acid (DNA) strands of chromosomes present in nuclei. The mechanisms by which information is extracted from DNA, processed, and translated into complete proteins collectively result in lean tissue growth. If one appreciates the extent to which lean tissues of the body increase during growth, and that the bulk of this increase is protein, then the importance of protein synthesis is realized.

2. Development of Carcass Tissues

At birth, nearly all structural tissue cell have been developed, and therefore bone and muscle cells only increase in size (hypertrophy) rather than number (hyperplasia). Due to excessive calorie during growth, fat cells may increase in number. While fat tissue grows during all stages of development, structural tissue growth eventually ceases (Figure 3.2). Bone is the first to stop growing. Primary bone development occurs during the early phase of postnatal growth, and the length of this period of growth is dependant on how tall and long the individual animal will be at maturity. For example, a moderately sized heifer may be near her mature height as a yearling, whereas, a large framed heifer may continue to grow in height and length until 24 months of age, and a small framed, quick patterned heifer may have reached her mature size at 8 months. Regardless of when this occurs chronologically, the exponential development of muscle and fat tissue does not occur until bone growth begins to wane.

Once muscle development begins to increase substantially, it will continue to do so until hypertrophy nears completion given an animal has sufficient nutrient availability. As the increase in cellular muscle tissue size reaches a maximum, fat (adipose) tissue then becomes the greater in development. The goal is for an animal to reach it's mature size, maximize it's muscle content, and have an acceptable amount of fat to happen at an optimal endpoint where both the livestock producer and the packer are able to generate a profit. This point is known as the point of inhibition, time of the growth curve when muscle development slows, and fat development increases to a point when it is no longer

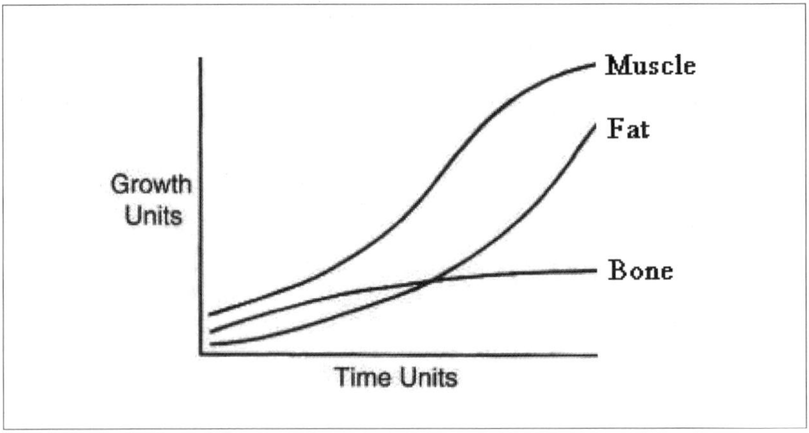

Figure 3.2. Postnatal growth curves of muscle, fat and bone.

profitable for an animal to be on feed.

In the livestock industry, the first role to determine whether an animal has enough muscle content to maximize their endpoint value at an acceptable weight and percent yield, must therefore to understand the way adipose tissue develops. Adipose tissue is divided into four sub areas based on the way it is deposited by the body. The four area are:

1) Subcutaneous – immediately under the skin; external fat,

2) Intermuscular – in between muscle tissues; seam fat,

3) Intramuscular – within the muscle tissues; marbling,

4) Kidney – the fat that surrounds the kidney and pelvic region, as well as the heart (KPH).

Subcutaneous fat is the only type of fat deposit that can be evaluated (without ultrasonography) visually, and is measured on the carcasses of all meat animal species. The measurement for subcutaneous fat is obtained from the back of carcasses, and is therefore referred to as backfat. When ribbed (opened for taken carcass measurements), sheep and cattle carcasses are ribbed between the 12th and 13th ribs (final ribs). Accordingly, backfat is measured at the 12th rib of sheep and cattle. However, swine carcasses are ribbed at the 10th rib, and therefore backfat is measured at that location. Subcutaneous fat is often referred to as external fat because it is on the external portion of the carcass and this type of fat is easy to see in live animals.

Intermuscular fat has been termed "seam fat", although it is different in the seam fat in swine. Seam fat in swine refers to the subcutaneous fat visible between the back legs at the base or "seam" of the ham. Intermuscular seam fat is the fat that develops between muscles. It is difficult to see in live animals, and a practical way of measuring seam fat has not been established in carcasses.

Intramuscular fat is highly important in determining meat grade. The amount of intramuscular fat in a carcass is referred to as marbling. While marbling is present in all species, it is critical in beef cattle because the quality grade is based in part on marbling (maturity is the other factor; physiological age). The intramuscular fat that is evaluated in sheep carcasses is located near the flank and is called "flank streaking". In swine, it is evaluated on the inside of the ribs, and is called "feathering".

Kidney fat is the fat deposited in the abdominal cavity. In sheep and swine, cannot evaluate kidney fat as it is removed with the internal organs. Beef carcasses, however, have the kidneys left intact, and this fat in addition to the fat surrounding the pelvic region and the vital organs of the forequarter (heart fat) are evaluated for yield grade calculation purposes. This estimate is termed KPH, and is represented as a percentage of body weight.

3. Skeletal Muscle Development

Skeletal muscle is a complex and heterogeneous tissue, and considered as an organ for the muscular system which consisting of the tissues of skeletal muscle, connective, nerve and blood or vascular to serve multitude of functions. It derives from paraxial mesoderm and divides into somites and notochord in vertebrate animals like mammals and avian species. This process is illustrated in Figure 3.3, beginning with the fertilized ovum and extending to the major tissues of the carcass. The fetal stage not only involves the development of muscle cells

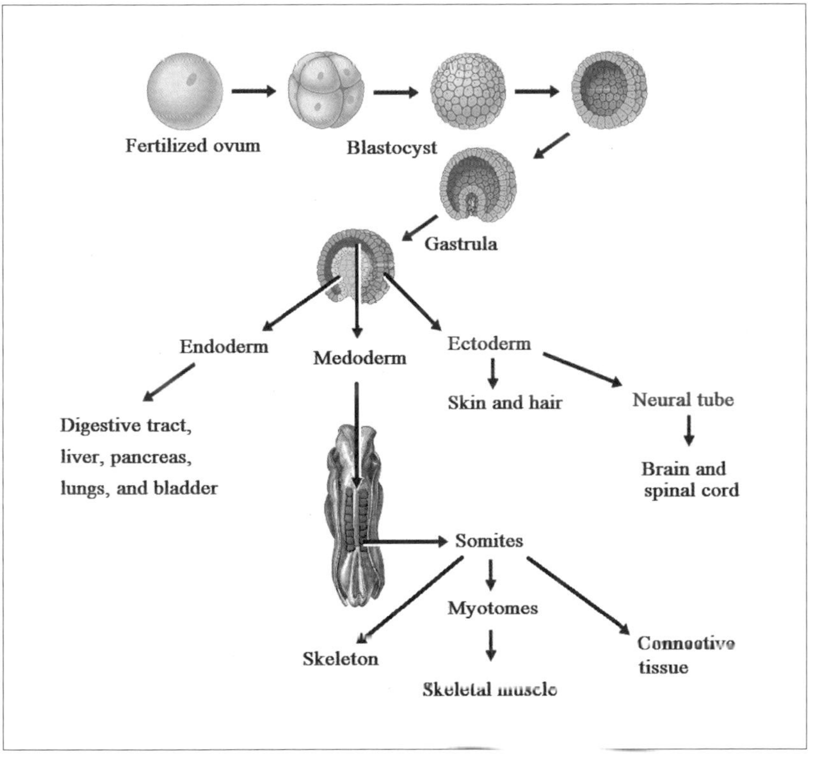

Figure 3.3. The progressive differentiation of tissues during prenatal growth and development.

(myogenesis) but also adipocyte and fibroblast of animals. In adults period, skeletal muscles are composed of bundles of multinucleated myofibers which distributed tendon to tendon and provided contractile activity in skeletal muscle.

3.1. Origin of Skeletal Muscle

The initial embryonic patterns can be classified into the ectoderm, mesoderm and endoderm. In vertebrate, the formation of skeletal muscle is mainly derived from the paraxial mesoderm during embryogenesis and the accumulation of paraxial mesoderm forms

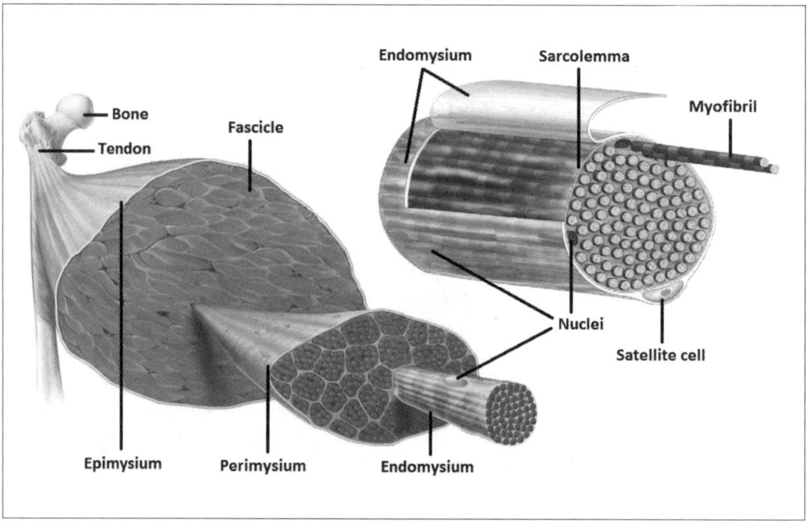

Figure 3.4. The fundamental structure of skeletal muscle made up from a number of bundled muscle fibers. The myofibrils contain contractile unit (or sarcomere) and also surrounded by a basal lamina under which satellite cells are located

segmented epithelial spheres which referred to somites. The first phase of skeletal formation in mammals, poultry and fish is primary myotome which originates from the four epithelial border of the dermomyotome. This dermomyotome is actually from somite after undergoing differentiation and subdivide and after that forms dorsal dermomyotome and ventral sclerotome. But the limb muscles vertebrate originate from muscle progenitors which delaminate from the ventrolateral border of the dermomyotome.

The fundamental structure of skeletal muscle includes myocytes, adipocytes and fibroblasts. These cells are derived from the same mesenchymal stem cell in fetal muscle of farm animals. This skeletal muscle formation begins during embryonic life and the process is completed after postnatal growth when the animal is turned to adult stage. During the development of skeletal muscle, thousands of cylindrical multinucleated muscle fibers is formed composing an array of stacked myofibrils running the entire length of the cell (Figure 3.4). Muscle mass and its multiplication are mainly depend on the number and size of fibers, also genetic and environmental factors.

3.2. Myogenesis

Skeletal myogenesis is initiated by the generation of myoblasts followed by their differentiation to myocytes and the formation of myofibers. The timing of different developmental stage varied significantly but the end products are the same to form multinucleated myofibers with contractile capability. Although alterations in growth and development of muscle can affect meat quality, they will not be covered in great detail in this text. Rather an only understanding of growth processes important to meat scientists will be covered.

The key factors of muscle growth are classified into prenatal stage (proliferation and differentiation of myogenic cells) and postnatal stage (satellite cells). This formation is quite unique because muscle cells of muscle fibers are multinucleated. The growth of muscle tissue is accomplished two fundamental biological processes which are protein accretion and cell proliferation and finally produce meat. First, myogenic cells are differentiated and undergo proliferative mitosis until further differentiation occurs into myoblasts. These myoblasts produce during myogenesis as a result of the final quantal cell cycle which are no longer capable to proliferate further, and therefore permanent fuse and synthesize myofibrillar proteins from the cell cycle. Myoblasts are cells that have the ability to fuse with one another to form immature muscle fibers. Thus, groups of myoblasts (muscle precursor cells) fuse and donate their nuclei to a single immature muscle fiber. Embryonic development of skeletal muscle fibers is shown in Figure 3.5.

Muscle fibers are produced from the accumulation of mature myotube derived myofibrils. In early myogenesis, myoblasts which are developed from mesodermal myogenic precursor cells are proliferated and differentiated into post-mitotic myocytes and lastly fuse to form a multinucleated myotube. Furthermore, the first myotubes formed in developing embryos and fetuses are known as primary myotubes and the muscle fibers they produce are primary muscle fiber (Figure 3.6). Secondary myotubes, also the result of myoblast fusion, form later in fetal development around the primary myotubes and produce secondary muscle fibers. As both types of fibers mature and begin to accumulate contractile protein, myofibrils are formed centrally

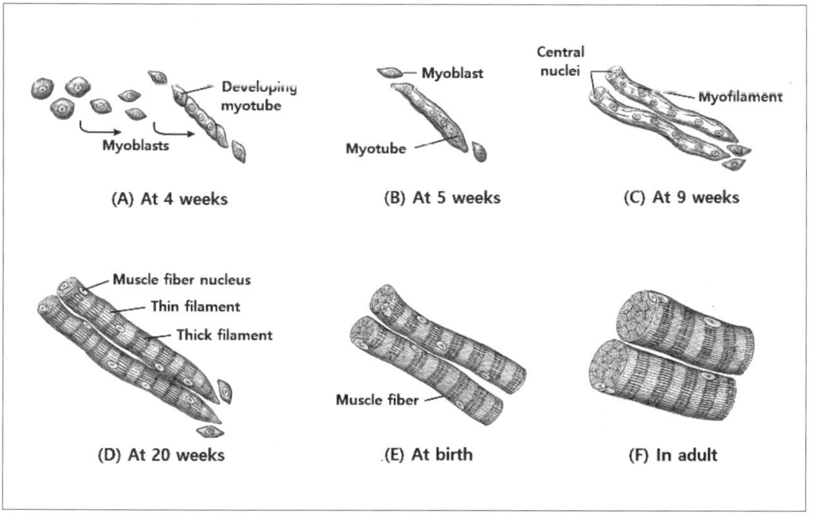

Figure 3.5. Figure 3.5. Embryonic development of skeletal muscle fibers. (A) Mesenchymal cells differentiate into myoblasts, which begin to form aggregates and line up into rows. (B) Myotubes lengthen by incorporating additional myoblsts via cell fusion. (C) Myofilaments have appeared but nuclei are still centralized (D) Developing muscle fibers now have cross-striated appearance and nuclei move peripherally. (E) Myofibrils have formed and nuclei have now shifted to periphery. (F) Muscle fibers are now thick and mature and consist of alternate thick and thin myofilaments grouped into longitudinal bundles as myfibrils with nuclei located at periphery (satellite cells are not shown).

in the muscle fiber, forcing the nuclei to the periphery. At this point, both primary and secondary muscle fibers structurally resemble mature muscle fibers. For any deficiencies, the appropriate number of embryonic muscle cells is produced that unlikely to be compensated during later development of muscle. The study on double-muscled cattle in prenatal period found that delaying differentiation and extending myoblast proliferation result higher number of muscle fibers.

Postnatal muscle growth occurs mainly for the increase of muscle fiber size instead of forming new fiber. Muscle growth in this time is considered due to muscle cell hypertrophy as muscle fiber hyperplasia ceases during prenatal period. During the last one-third of the prenatal period, hypertrophy (increase in size of existing fibers) contributes comparatively more to muscle growth that it does during the initial two-thirds. Developing muscle fibers grow in length by the addition of complete sarcomere units to the ends of existing myofibrils.

Satellite cells are the likely myogenic precursors for muscle regeneration, and they are seen to undergo cell division in regenerating muscle. Muscle regeneration requires the recruitment of an undifferentiated progenitor from the injured site to execute this process by satellite cells. Satellite cells reside in a niche on the surface of the muscle fiber, beneath the ensheathing basal lamina (Figure 3.7 and Figure 3.4), and are found in a similar location in many vertebrate species. Satellite cells not only generate myoblasts, but also re-appear in their niche as myotubes reform. Thus, satellite cells not only generate

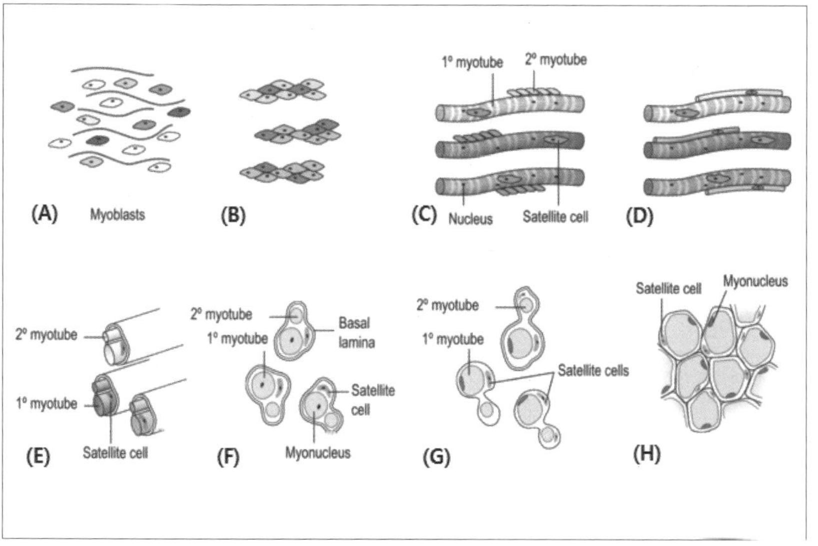

Figure 3.6. Schematic diagram of skeletal muscle development. (A) Primitive cells differentiate into myoblasts. (B) Myoblasts fuse together to form primary myotubes. (C–F) Later, secondary myotubes arise beneath the basal lamina of the primary myotubes. Fusion of myoblasts radially and longitudinally results in the formation of the muscle fiber beneath the basal lamina (shown in cross-section in E through H). In addition, some unfused myoblasts remain as satellite cells, which are maintained in the mature tissue. (G) As the muscle matures, primary and secondary myotubes separate, each with myonuclei and satellite cells, to become a mature fiber. (H) Finally, as the muscle fibers grow, they become arranged as tightly packed polygonal cells, characteristic of adult muscle.

differentiated progeny, but also maintain their own population by self-renewal, they can be classified as myogenic stem cells (monopotent muscle-resident myogenic stem cells).

3.3. Muscle Fiber Hypertrophy

Muscle hypertrophy can be considered distinct from muscle hyperplasia. During hypertrophy, contractile elements enlarge and the extracellular matrix expands to support growth. However hyperplasia in contrast which results in an increase in the number of fibers within a muscle. This hypertrophy has the greatest increase both their diameter and length after birth, but the rate of increase declines as animals approach maturity. In some mammals the number of muscle fibers does not increase appreciably after birth, and the muscle growth during postnatal is accomplished primarily through hypertrophy of existing muscle fibers.

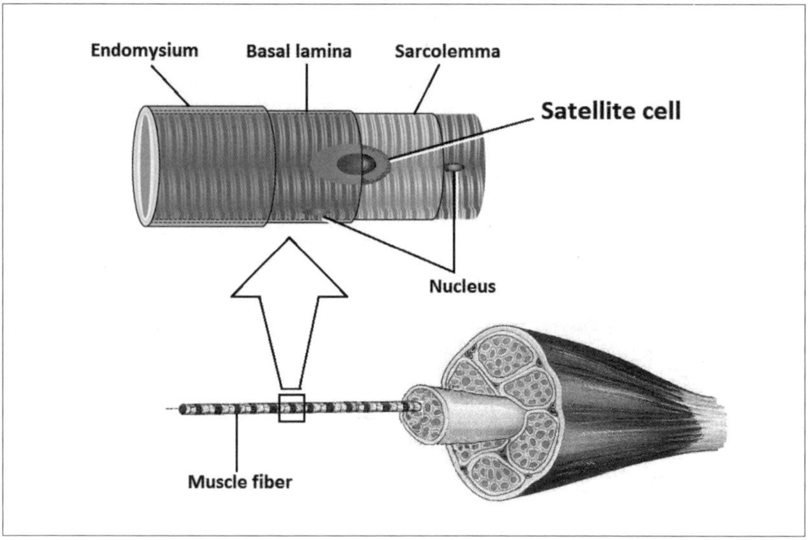

Figure 3.7. Muscle structure and the satellite cell niche

Hypertrophy is thought to be mediated by the activity of satellite cells. These cells become active when a sufficient stimulus is imposed on skeletal muscle. Once aroused, satellite cells proliferate and ultimately fuse to existing cells or among themselves to create new myofibers, providing the precursors needed for repair and subsequent growth of new muscle tissue. In addition, satellite cells are thought to facilitate muscle hypertrophy in several ways. For one, they donate extra nuclei to muscle fibers, increasing the capacity to synthesize new contractile proteins. Because a muscle's nuclear-content-to-fiber mass ratio remains constant during hypertrophy, changes require an external source of mitotically active cells. Satellite cells retain mitotic capability and thus serve as the pool of a myonuclei to support muscle growth. This is consistent with the concept of myonuclear domain, which proposes that the myonucleus regulates mRNA production for a finite sarcoplasmic volume and any increases in fiber size must be accompanied by a proportional increase in myonuclei. Given that muscles are comprised of multiple myonuclear domains, hypertrophy could conceivably occur as a result of either an increase in the number of domains (via an increase in myonuclear number) or an increase in the size of existing domains. Both are thought to occur in hypertrophy, with a significant contribution from satellite cells.

3.4. Development of Connective Tissue

Connective tissue is mainly synthesized by intramuscular fibroblast.

Fibroblast is derived from a common pool of progenitor cells together with myocytes and adipocytes during the early embryonic development. During early skeletal muscle development, mesenchymal stem cells first diverge into either myogenic or non-myogenic lineages. Myogenic mesenchymal progenitors develop into muscle fibers and satellite cells, while non-myogenic progenitor cells then develop into the stromal-vascular fraction of skeletal muscle wherein adipocytes, fibroblasts and resident mesenchymal progenitor cells. These non-myogenic progenitors have adipogenic and fibrogenic capacity and both of them are undergo differentiation and proliferation, leading to the elevation of both marbling and connective tissue content in the resulting meat products. Furthermore, by enhancing their conversion to adipogenesis reduces fibrogenesis, which likely results in the overall improvement of marbling (more intramuscular adipocytes) and tenderness (less connective tissue) of meat.

Muscle regeneration involves extensive proliferation and myogenic differentiation of satellite cells. Shortly after muscle injury, both satellite cells and non-myogenic progenitor cells are activated and proliferate; consequences, facilitate muscle regeneration. In addition, intramuscular fibroblasts particularly promote slow myogenesis, thus affecting muscle fiber type composition and overall maturation during muscle development. Extracellular component, collagen VI, regulates satellite cell self-renewal and differentiation. Besides, other components of extracellular matrix, such as proteoglycan, regulate proliferation and differentiation of satellite cells. Decorin, a small leucine-rich proteoglycan, traps transforming growth factor (TGF)-β to regulate satellite cell activation and muscle growth. Fibrogenesis is mainly regulated by the TGF-β signaling pathway, and a number of factors affect connective tissue deposition via altering TGF-β signaling. Extracellular matrix, apart of the intramuscular connective tissue, provides a niche environment to regulate myogenic differentiation of satellite cells and muscle growth.

3.5. Muscle Fiber Type Development

In general, muscle fiber types in adult skeletal muscle can be categorized as type I (slow oxidative), type IIA (fast oxido-glycolytic), type IIB (fast glycolytic). These muscle fibers represent a relative heterogenous population differing by its speed of shortening and by its metabolism. In detail, type I is low ATPase activity in alkali pH, very oxidative and is not glycolytic. Activity of type I fibers is labile at alkali pH and resistant at acidic pH. Whereas, type II is high ATPase activity in alkali pH, very glycolytic and is not oxidative. On the contrary, ATPase activity of type II fibers is alkali-resistant and labile to acid. Further classification of type II fibers are IIA which ATPase activity inhibited after preincubation at $pH<4.9$, type IIB which ATPase activity inhibited after preincubation at $pH<4.3$, and type IIC is partially resistant at this last pH.

As discussed in previous part, muscle fibers are multinucleated cells, from myoblasts (mononucleated cells) to form myotubes which then differentiate into fibers. The precursors of muscle fibers are derived from paraxial mesoderm. Myogenic specification in the embryo results in diverse fiber types and these diversity could be generated by having several populations of myogenic cells each programmed to produce particular development and physiological types of fiber. These can be associated to somitic, embryonic, foetal, and satellite (or adult) myoblasts, in accordance with ages from which they have been isolated. In some species such as chicken, rodents, and big mammals, the myogenesis is a biphasic process involving at least two sequentially generated populations of myotubes.

As first stage myotube (primary myotubes) population appears early and next population (secondary myotubes) appears larger than first one. The diameter of primary myotubes increase two-fold prior to onset of secondary myogenesis. These suggested that different populations of myotubes are originated from different generations of myoblasts in which produce the primary and secondary muscle fibers. These two populations of myoblasts varied in the isoforms of myosin heavy chain that they expressed. These sequences in MHC isoforms transition in myoblasts and myotubes are important to determine muscle fiber function in fetal, neonatal and adult stages.

Generally, slow fibers are generated largely from the first generation and the few fibers exist from the second generation in slow muscles, and normally fast fibers are generated from the second generation. In bovine, primary myotubes were observed from 30 days of gestation. In the beginning they contain MHC slow, fast, MHC-emb, MHC-fet, they differentiate rapidly and contain only MHC slow form 200 days of gestation. They produce type I fibers in all muscles. Secondary myotubes are observed at approximately 90 days, and they also contain the same as in 30 days gestation, MHC slow, fast, MHC-emb and MHC-fet. But then they develop into type II fibers (IIA, IIB, IIC) and produce a few number of type I fibers. The implication of a third generation corresponding to very small fibers considered as secondary fibers. This is because the existence of a third generation was described in several species such as rat, sheep and human.

The contractile differentiation is practically complete in newborn animals, for instance human and bovine, that are relatively mature at birth which adult MHC isoforms appear during the end of gestation. In the prenatal period is characterized by the decrease in the proportion of type IIC fibers which differentiate in type I or II, depending on the muscle. Contrary, in species less mature at birth (pig), muscle fiber type I percentage increase during the prenatal stage, and at this stage the disappearance of MHC-fet and MHC-emb. However the postnatal changes in fiber types vary between species. In mammals, an increase in type IIB fibers was observed during growth as detriment of type IIA. It seems that this conversion begins soon after birth particularly in cattle. The increase in type IIB is accompanied by an increase of type IIX (IID) fibers. Also, the development of muscle fiber differs between muscle.

4. Growth and Development of Adipose Tissue

Adipose tissue is a connective tissue that is derived from the mesoderm of the developing embryo. The mesenchymal cells give rise to an early form of fat cells knows as adipoblasts (Figure 3.8). These adipoblasts formed from adipose tissue that is gradually developed into lobes and lobules which are encased in a delicate sheath of collagenous fibers. Adipoblasts will continue to proliferate until a variety of cells send signals to stop replicating. At this point, adipoblasts may differentiate into preadipocytes. As lipid droplets form in the preadipocyte, they merge to become one large globule which is characteristic of a mature

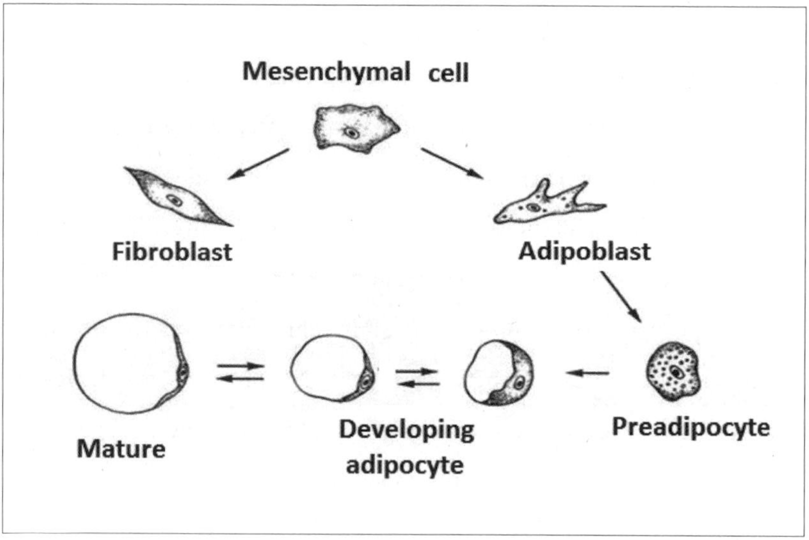

Figure 3.8. Diagrammatic representation of adipocyte development

adipocyte. At this stage, the cytoplasm, nucleus, mitochondria, and other organelles are located in one side as it becomes filled with lipid. Mature adipocytes are the predominant cell type in adipose tissue and the mature adipocyte is expected to have a diameter around 120 μm which is larger than primitive cell diameter size at 1 to 2 μm.

The development of adipoblasts are occurred at different rate in different parts of the body, resulted to the variations proportions of fat present in the body. In mammals, there are two main classes of adipose tissue that are white adipose tissue (WAT) and brown adipose tissue (BAT). WAT is designed for energy storage whereas BAT is to dissipates energy and generates heat. WAT is far more prevalent and more biologically relevant, whereas BAT primarily functions in the neonatal period. White adipose tissues coordinate systemic metabolism which is highlighted the key roles of metabolic dysfunctions (e.g. hyperglycemia, hyperlipidemia, hypertension, diabetes, liver disease, increase carcinogenesis, etc.) that are associated to too many (obesity) or too few (lipodystrophy) white adipocytes. Brown adipocytes convert nutrients into chemical energy in the form of heat. The main role of BAT in neonatal period is to challenge the temperature control. This is because brown fat cells express a unique thermogenic and mitochondrial genetic program that promotes mitochondrial biogenesis, energy uncoupling and energy dissipation, which provides essential heat to the organism.

In young animals, deposits of fat usually appear first in visceral areas. Then, if nutrient intake is adequate, fat is deposited beneath the skin (subcutaneous), between muscles (intermuscular), and between muscle bundles (intramuscular). Intramuscular fat deposits are the latest to develop and occur in fat cells associated with perimysial connective tissues. These fat cells accumulate and grow between the bundles of muscle fibers in close proximity to the arterioles that permeate the muscle. The enlargement of adipose tissue within an animal can occur by hyperplasia or hypertrophy. As an animal grows and fat is deposited, a population of cells accumulate lipid and increase in volume and diameter. The increase in adipocyte number also can be probably associated to the preadipocytes filling with lipid and becoming large.

Marbling adipose tissue, also known as interfascicular or intramuscular adipose tissue, represent a unique adipose tissue depot. In cattle as well as other livestock species, it can be distinguished from other fat depots by its location within perimyisal connective tissues alongside muscle fibers (Figure 3.9A and 3.9B). Marbling consists of clusters of individual cells (adipocytes), and intramuscular adipocytes increase both in number and size in cattle as beef progresses from being practically devoid to higher marbling scores. In pectoralis muscle of chickens that were feed-restricted immediately post hatch, marbling adipocytes formed between muscle bundles as well as within muscle bundles (Figure 3.10). As compared to cattle, the deposition of adipocytes within muscle bundles in chicken muscle is more consistent with the conversion of muscle satellite cells to adipocytes. The

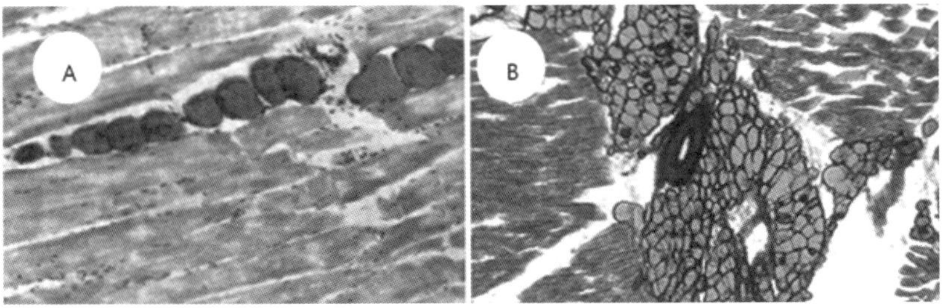

Figure 3.9. A: Marbling adipocytes lying alongside muscle fibers in bovine longissimus muscle, B: Marbling adipocytes located in perimysial seams of connective tissue. The oval central structure is an arteriole, which is flanked by pockets of marbling adipocytes.

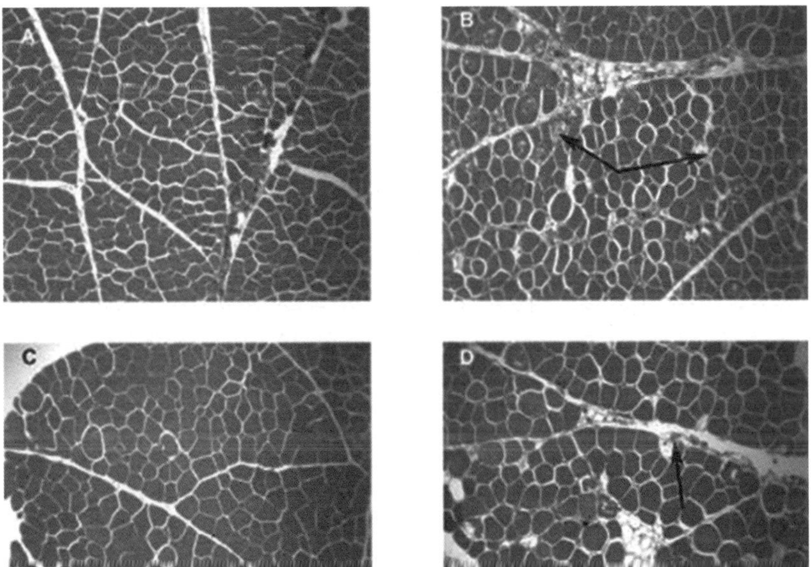

Figure 3.10. Morphological structure of the pectoralis muscle at 35 and 42 days of age in control (A and C0 and growth-restricted chickens (B and D). The muscle samples were taken at 35 (A and B) and 42 days of age (C and D). The arrows highlight muscle fibers undergoing lysis..

mechanisms responsible for the trans-differentiation of muscle cells to adipocytes in beef cattle still have not been identified as this mechanism of changes is quite difficult to measure in livestock species. However, cell culture studies suggest that fatty acids derived from marbling adipocytes within muscles actually may promote this process.

5. Growth and Development of Bone

Bone is a mineralized and vascularized tissue which is the main component of vertebrates skeleton, and supports the vertebrate body. Formation of bone is a continuous process during embryonic and postnatal stage. Bones in vertebrates are formed via two different processes: intramembranous and endochondral ossification (Figure 3.11). In intramembranous ossification, osseous tissues is produced from mesenchymal progenitors, this process generates the bone collar and sequentially increases in its diameter, in the case of long bones in mammals. Contrary, endochondral ossification the cartilaginous mold is replaced by the bone tissues, and is essential for epiphyseal morphogenesis and longitudinal growth of long bones.

During long bone development mesenchymal cells aggregate to form a precartilage condensation, which is a primordium of a long bone. From the center of the primordium, mesenchymal cells differentiate into chondrocytes and start to produce cartilaginous extracellular matrix (ECM). Condensation proceeds particularly

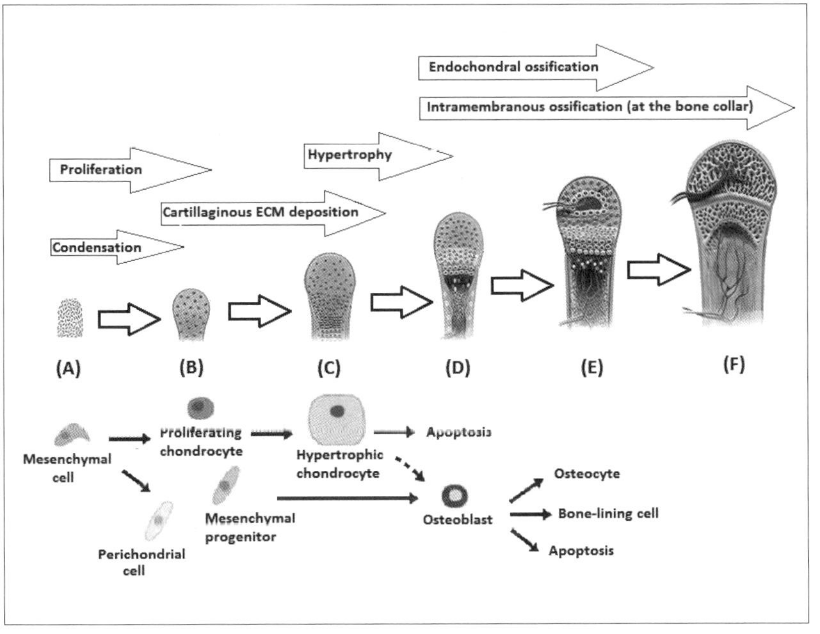

Figure 3.11. An illustration of the development of a long bone. (A) Mesenchymal cells aggregate to form a precartilage condensation, which is a primordium of a long bone. (B) Chondrocytes start to produce cartilaginous extracellular matrix (ECM). (C) Hypertrophic chondrocytes formed. (D) Formation of bone collar via intramembranous ossification. (E) Endochondral ossification starts in both epiphyses, the perichondrium disappears from both articular ends, and the articular cartilage begins to differentiate. (F) The epiphyseal growth plate persists.

around the presumptive epiphysis region. At the periphery of the cartilaginous anlage, mesenchymal cells begin to differentiate into perichondrial cells. Chondrocytes undergo further maturation and become hypertrophic chondrocytes from the center of the cartilaginous anlage. The whole skeletal primordium including the articular surface becomes surrounded by the perichondrium, which eventually becomes bi-layered at the diaphyseal region. Some perichondrial cells in the inner layer of perichondrium differentiate into osteoblasts and form the bone collar via intramembranous ossification. The zone of hypertrophic chondrocytes becomes vascularized, and perichondrial cells migrate into the anlage and differentiate into osteoblasts. Hypertrophic chondrocytes undergo apoptotic cell death and are replaced by osteoblasts, which subsequently form trabecular bones (endochondral ossification), resulting in formation of the marrow cavity. A small population of hypertrophic chondrocytes in a particular region are shown to differentiate into osteoblast-like cells; however, those cells are not a major source of osteoblasts (indicated by a dashed arrow in Figure 3.11). Proliferating, prehypertrophic and hypertrophic chondrocytes are arranged in columns. This chondrogenic zone is called the growth plate. Endochondral ossification spreads outward. The perichondrium transforms to the periosteum along the bone collar. Eventually, endochondral ossification starts also in both epiphyses. Around this period, the perichondrium disappears from both articular ends, and the articular cartilage begins to differentiate.

After development, bone is continuously replaced throughout life whereby osteoclasts, osteoblasts and osteocytes interact and control resorption and replacement of old bone. Bone remodelling is a balance between formation by osteoblasts and resorption by osteoclasts. This could be accomplished by the basic multicellular unit, which contain osteoblasts, osteoclasts, blood vessel, nerves, and connective tissue. In the basic multicellular units, osteoblasts are directly active in production of bone and are responsible for bone deposition in the various sites of bone development, while osteocytes help to maintain it. Concurrently with bone growth in length and diameter, the marrow cavity becomes enlarged due to the continuous resorption of the medullary wall by activity of osteoclasts. This cavity is occupied by marrow, a soft tissue composed of networks of delicate reticular connective tissue filled with various cells. Two types of marrow are recognized: red and yellow. Red marrow is the chief blood cell forming organ if the adult body and is found in spongy bone of the epiphysis of long bones, and the main bodies of the sternum, ribs, and vertebrae. Yellow marrow consists mainly of adipose tissue and is found in the medullary cavity of bones. In the adult animal, the site of the original diaphysis has been replaced by the medullary cavity, and the bone shaft has been derived from intramembranous growth of bone beneath the periosteum.

Chapter 4.

Muscle Contraction Mechanism

Living muscle is a highly specialized tissue that is capable of converting chemical energy into mechanical energy through its contraction. To understand a mechanism of muscle contraction, it is necessary to outline the structure of skeletal muscle, which is surprisingly orderly and analogous. The overall structure of muscle is designed for contraction and relaxation, which leads to movement and locomotion. This ability to contract and relax lost during the transformation of muscle to meat. Yet, certain aspects of contraction and relaxation in the living muscle relate directly to the shortening and decreased tenderness that occurs in meat during this postmortem conversion. The events surrounding this whole process of muscle contraction and relaxation, and subsequent loss of these abilities, dramatically affect the palatability of meat. The biochemical processes that provide energy for muscle function in living animals are similar to those processes that cause an accumulation of metabolites and subsequent loss of water-holding capacity during the postmortem period. Thus, an understanding of how muscles function normally to produce movement and generate heat forms the basis for our understanding why they are important in determining the ultimate functional characteristics of meat as a source of food.

1. The Nature of Stimuli and Action Potentials

The sequence of events that result in the contraction of an individual muscle fiber begins with a signal from the motor neuron innervating that fiber. A nervous stimulus starts in the brain or spinal cord, and is transmitted to the muscle via nerves. Most contraction of skeletal muscles occur as a result of conscious effort originating in the brain. The brain sends an electrical stimulus through the nervous system to the motor neuron that innervates several muscle fibers. In the case of some reflexes, the signal to contract can originate in the spinal cord through a feedback loop with grey matter. Other actions such as locomotion, breathing, and chewing have a reflex aspect to them. The contractions can be initiated both consciously or unconsciously.

In living cells, under normal resting conditions, an electrical potential (difference in net charge) exists between the inside and outside of the cells. These potentials vary from 10 to 100 millivolts(mV), depending on the type of cell, but in resting (unstimulated) nerve and muscle fibers the potential is about 70 mV. The fluids that are inside and outside of these fibers contain approximately equal concentrations of positive and negative ions. Generally, a slight excess of negative ions accumulates in the intracellular fluid along the inner surface of the membrane, whereas a slight excess of positive ions is present along the extracellular surface of the membrane. As a result, an electrical potential (charge difference) or resting membrane potential is developed across the cell membrane. This membrane potential is positive on the outer surface and negative inside the cell (Figure 4.1).

Differences in the concentrations of ions on opposite sides of a cellular membrane lead to a voltage called the membrane potential. Typical values of membrane potential are in the range +40 mV to –70 mV. Many ions have a concentration gradient across the membrane, including potassium (K^+), which is at a high concentration inside and a low concentration outside the membrane. Sodium (Na^+) and chloride (Cl^-) ions are at high concentrations in the extracellular region, and low concentrations in the intracellular regions. These concentration gradients provide the potential energy to drive the formation of the membrane potential. This voltage is established when the membrane has permeability to one or more ions.

The membrane potential in nerve and muscle is the result of 1) the active transport of ions through the membrane, 2) the selective membrane permeability to ions and small molecules, and 3) the unique ionic composition of the intracellular and extracellular fluids. Extracellular fluid contains high concentrations of Na^+ and Cl^- ions, but very low concentrations of K^+ and nondiffusible negative ions. In contrast, concentrations of K^+ and nondiffusible negative ions are very

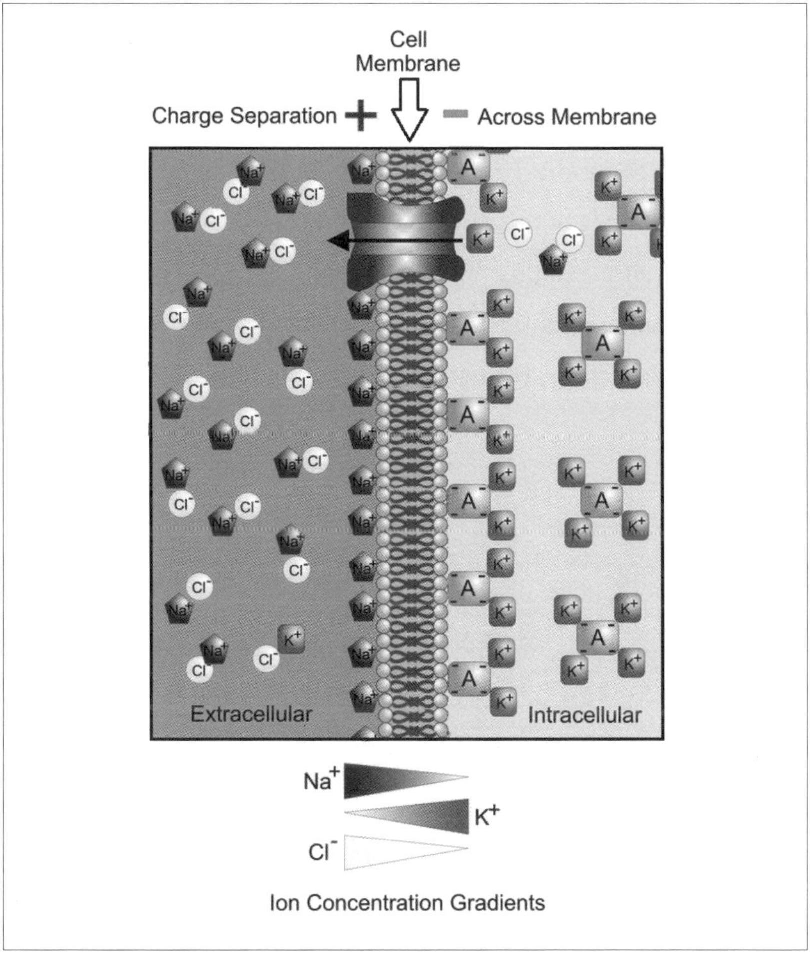

Figure 4.1 Establishment of a membrane potential in the normal resting nerve fiber, and the development of concentration differences of sodium (Na^+), potassium (K^+), chloride (Cl^-), and non-diffusible negative ions (A^-), between the two sides of the membrane. Rectangles are membrane-impermeable anions (these arise from a variety of sources including proteins). The large structure with an arrow represents a transmembrane potassium channel and the direction of net potassium movement.

high in the intracellular fluid while Na^+ and Cl^- concentrations are quite low. Concentration gradients of Na^+ and K^+ across transport of Na^+ out of the cell, and K^+ into the cell.

Two ions (Na^+ and K^+) are responsible for membrane potential. An unequal distribution of these two ions occurs on the two sides of a nerve and muscle cell membrane because carriers actively transport these two ions: sodium from the inside to the outside and potassium from the outside to the inside (Figure 4.2). As a result of this active transport

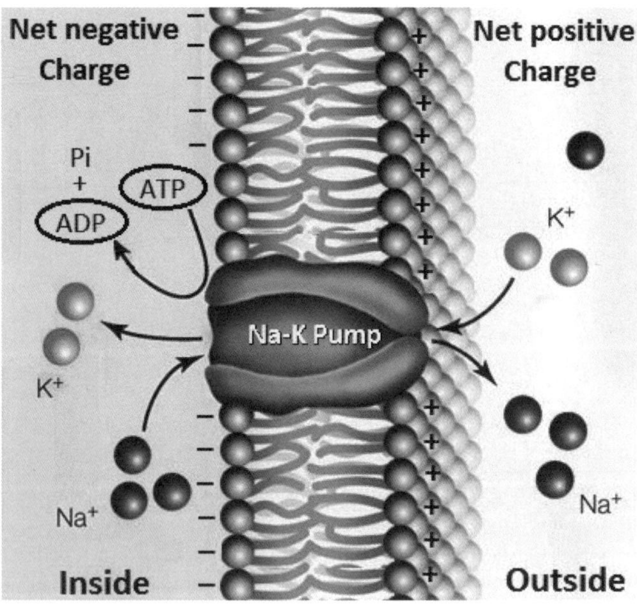

Figure 4.2. The sodium-potassium pump

mechanism (commonly referred to as the sodium-potassium pump), there is a higher concentration of sodium on the outside than the inside and a higher concentration of potassium on the inside than the outside.

Energy required to pump both types of ions across the membrane against a concentration gradient is furnished by adenosine triphosphate (ATP). Permeability of a plasma membrane is 50 to 100 times greater for the diffusion of K^+ than Na^+. Thus K^+ passes through the membrane with relative ease compared to Na^+. As a result, K^+ diffuse out of the cell more rapidly than Na^+ diffuse in. Of course, nondiffusible negative ions in the intracellular fluid pass through the membrane only with extreme difficulty.

Therefore, the net flow of electrical charges across the membrane is due to the diffusion of positively charged potassium ions into extracellular spaces, leaving fewer positive charges inside the cell. As a result, oppositely charged ions line the membrane, with the positive ions on the outer surface attracting negative ions on the inner surface of the membrane, thus establishing the resting membrane potential.

However, the net diffusion of positive charges out of the cell does not continue indefinitely. Once the resting membrane potential is established, further flow of K^+ out of the cell is impeded. Thus, an equilibrium is established between outward diffusion of K^+ (along its concentration gradient) and the force opposing outward diffusion: the positive potential established by the presence of cations on the outer surface of the plasma membrane.

Nerve and muscle fibers exhibit membrane potentials, as do other cells, but they have a unique capability not shared by any other cell type. They are able to transmit an electrical impulse, called an action potential, along their membrane surfaces. When an action potential is transferred from a motor nerve to muscle fibers, it initiates muscle contractions. An action potential travels along the membrane surface of the nerve fiber, and is actually a wave of reversing electrical charges or "depolarization" that results from chemical changes in the membrane. Thus, an action potential is often referred to as an electrochemical process.

As described in the previous section, in the resting state, the membrane is positive on its outer surface and negative on the inside. The action potential is initiated by a sudden and dramatic increase in the permeability of the membrane to Na^+. As a result, Na^+ rush into the cell to establish equilibrium between the concentrations inside and outside of the cell. At the same time, permeability of the membrane to K^+ remains the same and, as a result, an excess of positive charges enters the cell, reversing the resting membrane potential (Figure 4.3A).

An increased permeability to Na^+ lasts only for a small fraction of a millisecond. The newly reversed membrane potential reduces Na^+ permeability to its former low level. The outward flow of K^+ continues and re-establishes the resting membrane potential (Figure 4.3B). The sodium-potassium pump in the cell membrane redistributes Na^+ to the outside and K^+ to the inside without disrupting the membrane potential.

The entire sequence of events from initiation of an action potential to re-establishment of a resting membrane potential on the nerve fiber requires about 0.5 to 1 millisecond.

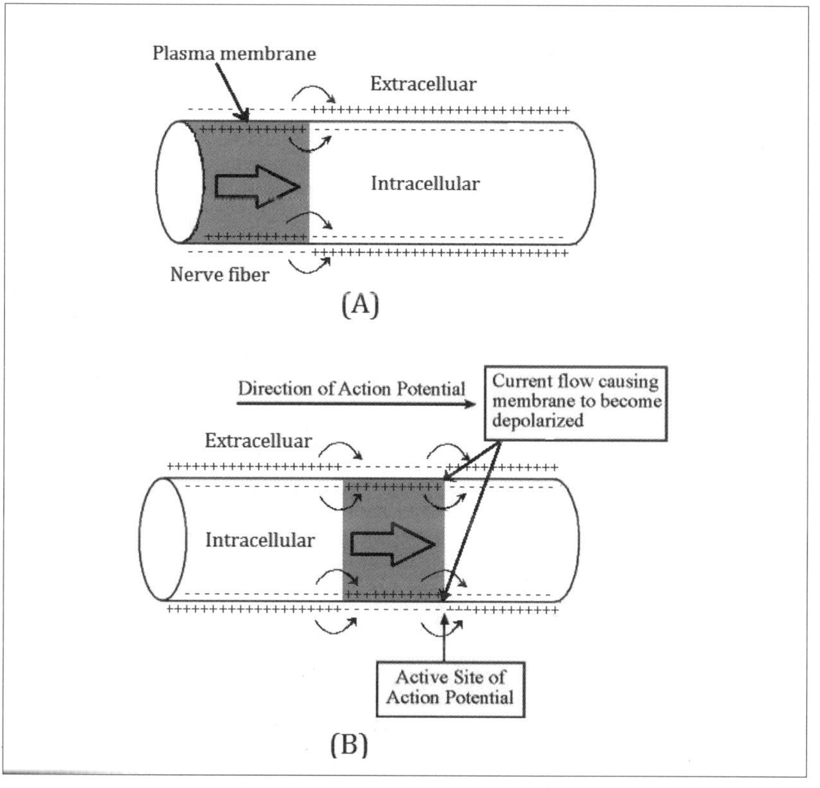

Figure 4.3. (A) The beginning of an action potential in a nerve fiber showing the reversed membrane potential. (B) The action potential, as it progresses along a nerve fiber, showing the depolarization at the front and repolarization at the rear. A large arrows indicate the direction of the action potential's movement. The curved arrows indicate points of reserving potential in the membrane.

2. Transmission at the Neuromuscular Junction

The neuromuscular junction is specialized on the nerve side and on the muscle side to transmit and receive chemical messages. Each motor neuron runs without interruption from the ventral horn of the spinal cord to the neuromuscular junction as a large, myelinated axon. As it approaches the muscle, it branches repeatedly to contact many muscle cells, to gather them into functional group known as a motor unit. As the terminal reaches the muscle fiber, it loses its myelin to form a

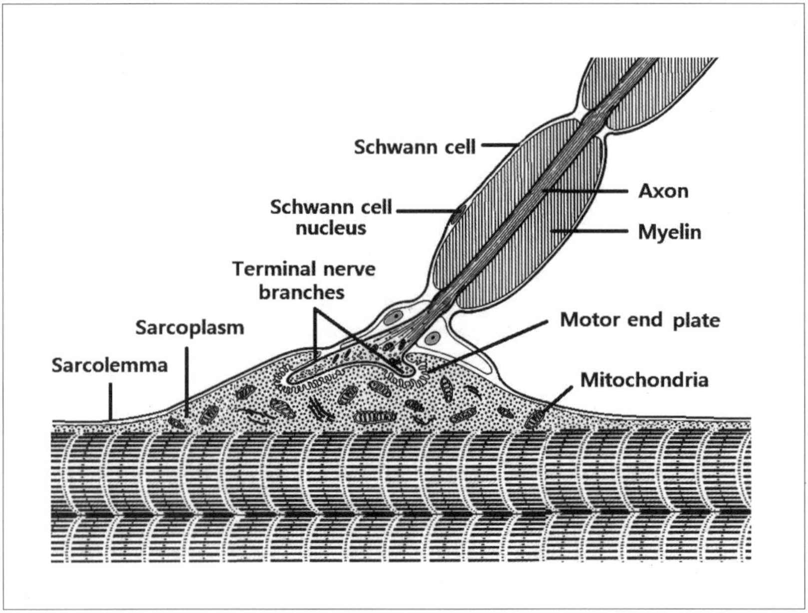

Figure 4.4. A schematic representation of the motor end plate, as seen in a histological section along the longitudinal axis of a muscle fiber.

spray of terminal branches against the muscle surface and is covered by Schwann cells. This arrangement conforms to the architecture on the synaptic area of muscle membrane (Figure 4.4). The nerve is separated from the surface of the muscle by a gap of approximately 20 nm, called the junctional cleft. The nerve and muscle are held in tight alignment by protein filaments called basal lamina, which span the cleft between nerve and end plate.

The axon of the motor nerve carries electrical signals from the spinal cord to the muscles and has all of the biochemical apparatus needed to transform the electrical signal into a chemical one. All the ion channels, enzymes, other proteins, macromolecules, and membrane components needed by the nerve ending to synthesize, store and release acethylcholine and other trophic factors are made in the cell body and are transmitted to the nerve ending by axonal transport (Figure 4.5). The simple molecule choline and acetate are obtained from the environment of the nerve ending, the former by a special system that transports it from the extracellular fluid to the cytoplasm and the latter in the form of acetyl coenzyme A from mitochondria. The enzyme choline acethyltransferase brings about the reaction of choline and acetate to form acetylcholine, which is stored in cytoplasm unit it is transported into vesicles, which are in a better position for release.

Acetylcholine, synthesized from choline and acetate by acetylcoenzyme A, is transported into coated vesicles which are moved to release sites. A presynaptic action potential, which triggers calcium influx through

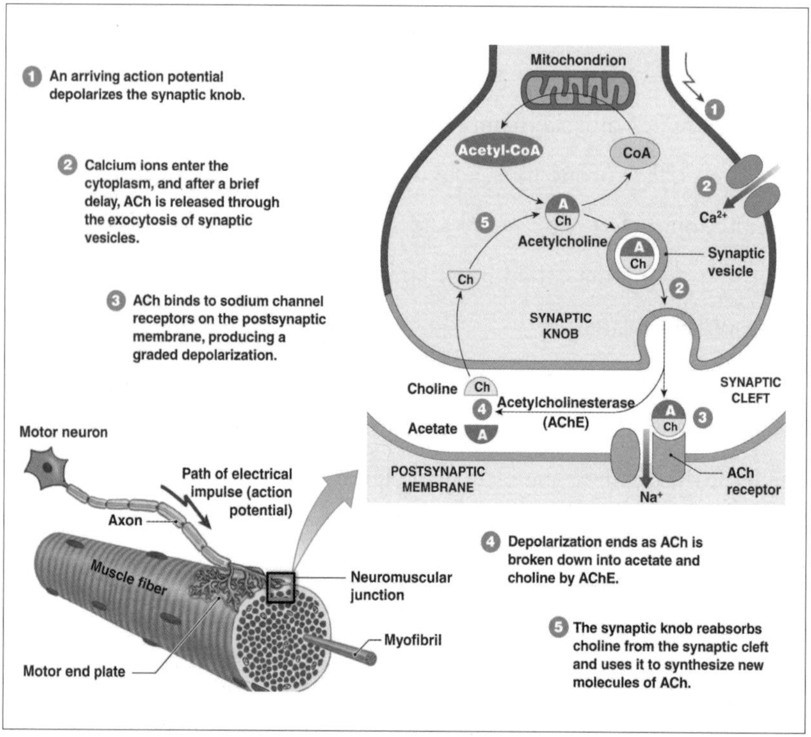

Figure 4.5. A schematic representation of the events that occur at a cholinergic synapse.

specialized proteins (Ca^{2+} channels), causes the vesicles to fuse with the membrane and discharge transmitter. Membrane from the vesicle is retracted from the nerve membrane and recycled. Each vesicle can undergo various degrees of release of contents-from incomplete to complete.

During a nerve action potential, sodium from outside flows across the membrane, and the resulting depolarizing voltage opens calcium channels, which allow entry of the calcium ions into the nerve and cause the release of acetylcholine. A nerve action potential is the normal activator that releases the transmitter acetylcholine. The number of quanta released by a stimulated nerve is greatly influenced by the concentration of ionized calcium in the extracellular fluid. If calcium is not present, depolarization of the nerve, even by electrical stimulation, will not produce release of transmitter. The calcium current persists until the membrane potential is returned to normal by outward fluxes of potassium from inside the nerve cell. With the calcium channels on the nerve terminal are the potassium channels, including the voltage-gated and calcium-activated potassium channels, whose function is to limit calcium entry into nerve and therefore depolarization.

3. Muscle Contraction Signals

In all muscle types, the contractile apparatus consists of two main proteins: actin and myosin. Striated muscle is so called because the regular arrangement of alternating actomyosin fibers gives it a striped appearance. This arrangement allows coordinated contraction of the whole muscle in response to neuronal stimulation through a voltage and calcium-dependent process known as excitation–contraction coupling. The coupling enables the rapid and coordinated contraction required of skeletal muscles and the heart.

In skeletal muscle cells, contraction thus depends on an increase

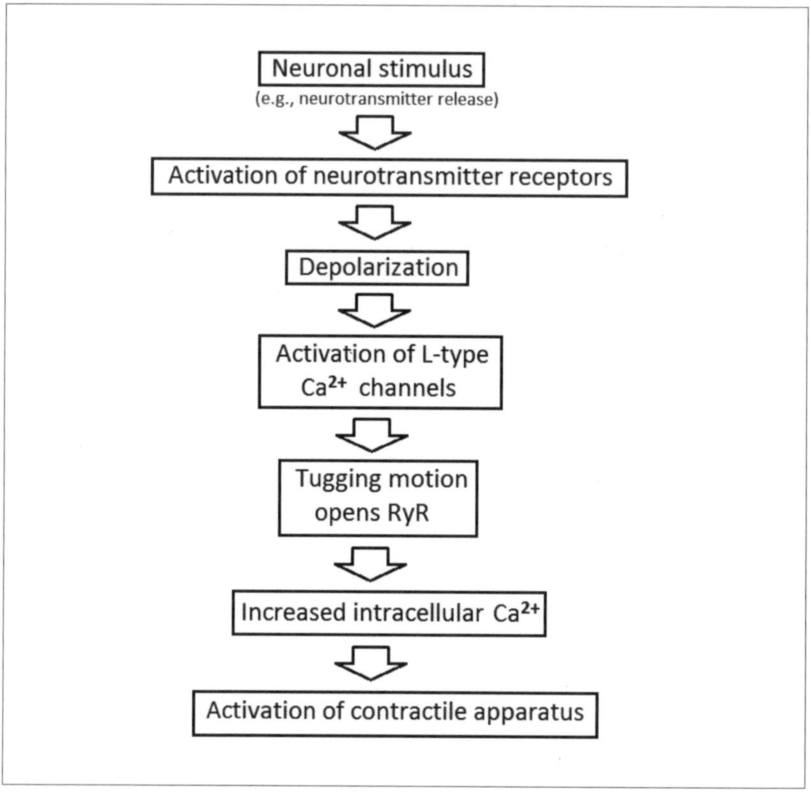

Figure 4.6. Overview of muscle contraction signals in skeletal muscle.

in cytosolic calcium concentration (Figure 4.6). Calcium has an extracellular concentration of 2~4 mM and a resting cytosolic concentration of 100 mM. It is also stored inside cells within the sarcoplasmic (SR) at a concentration of 0.4 mM. The increase in calcium levels is due to its release from the SR stores via ryanodine receptor (RyRs). Neurotransmitters such as acetylcholine bind to receptors on the muscle surface and elicit a depolarization by causing sodium/calcium ions to enter through associated channels. This shifts the resting membrane potential to a more positive value, which in turn activates voltage-gated channels, resulting in an action potential. The action potential stimulates L-type calcium channels (also known as dihydropyridine receptors). These are mechanically coupled to the SR RyRs and open them directly.

3.1 Calcium Release

Once intracellular calcium levels are raised, calcium binds to either troponin C on actin filaments (in striated muscle) or calmodulin (CaM), which regulates myosin filaments (in smooth muscle). In striated

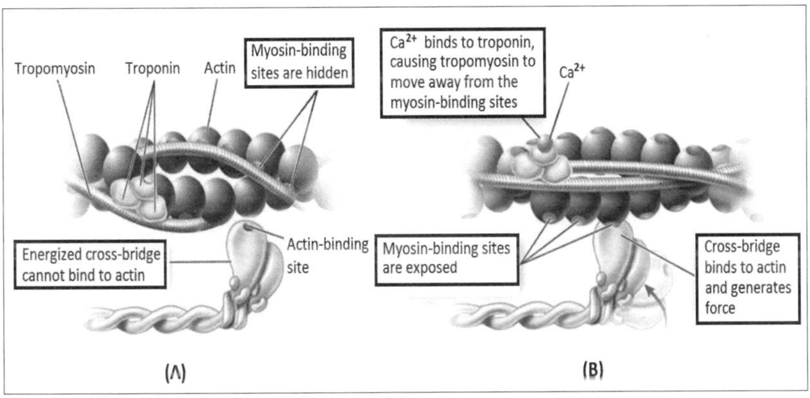

Figure 4.7. Calcium triggers contraction in skeletal muscle. (A) Muscle in the relaxed state has tropomyosin covering myosin-binding sites on actin. (B) Calcium binds to troponin C, which induces a conformational change in the troponin complex. This causes tropomyosin to move deeper into the actin groove, revealing the myosin-binding sites.

muscle, calcium causes a shift in the position of the troponin complex on actin filaments, which exposes myosin-binding sites (Figure 4.7). Myosin bound by ADP and inorganic phosphate (Pi) and then form cross-bridges with actin, and the release of ADP and Pi produces the power stroke that drives contraction. This force causes the thin actin filament to slide past the thick myosin filament and shortens the muscle. Binding of ATP to myosin then releases myosin from actin, and myosin hydrolyzes ATP to repeat the process (Figure 4.8).

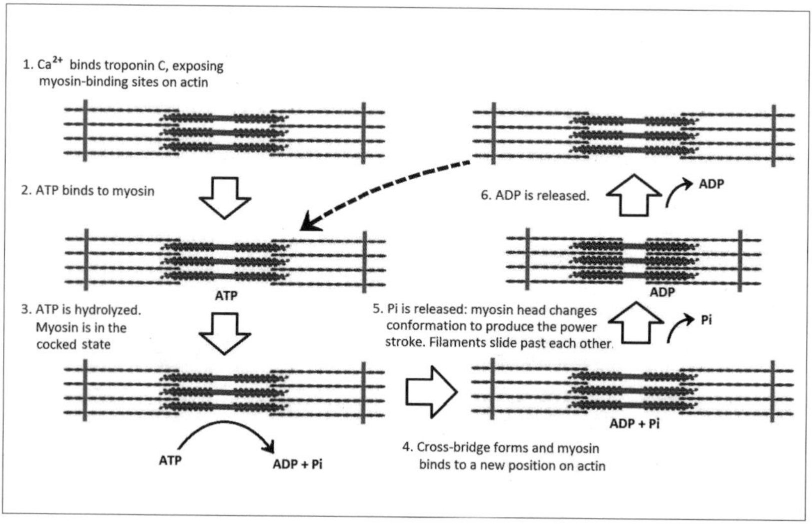

Figure 4.8. Cross-bridge cycle in striated muscle. (1) Calcium binds to troponin C, causing the conformational shift in tropomyosin that reveals myosin-binding sites on actin. (2) ATP then binds to myosin. (3) ATP is then hydrolyzed. (4) A cross-bridge forms and myosin binds to a new position on actin. (5) Pi is released and myosin changes conformation, resulting in the power stroke that causes the filaments to slide past each other. (6) ADP is then released.

3.2 Muscle Contraction

Skeletal muscles comprise multiple individual muscle fibers that are stimulated by motor neurons stemming from the spinal cord. They are grouped together to form "motor units" and more than one type of muscle fiber can be present within each motor unit. Muscle fibers can be divided into fast- and slow-twitch muscles. Fast-twitch muscles use glycolytic metabolism and are recruited for phasic activity (an active contraction). Slow-twitch muscles (also known as red muscles) are rich in myoglobin, mitochondria, and oxidative enzymes and specialized for sustained or tonic activity.

The neuromuscular junction (NMJ) that connects skeletal muscle with the nerves that innervate them consists of three distinct parts: the distal motor nerve ending, the synaptic cleft, and the postsynaptic region, located on the muscle membrane. Motor neurons branch into multiple termini, which are juxtaposed to motor end-plates, specialized regions of muscle where neurotransmitter receptors are concentrated (Figure 4.9). The transfer of information between the nerve and muscle is mediated by the release of acetylcholine from the motor neuron, which diffuses across the synaptic cleft, and binds to and activates the ligand-gated, nicotinic acetylcholine receptors (nAChRs) on the end-plate. Activation of the nAChR leads to an influx of cations (sodium and calcium) that causes depolarization of the muscle cell membrane. This depolarization in turn activates a high density of voltage-gated sodium channels on the muscle membrane, eliciting an action potential.

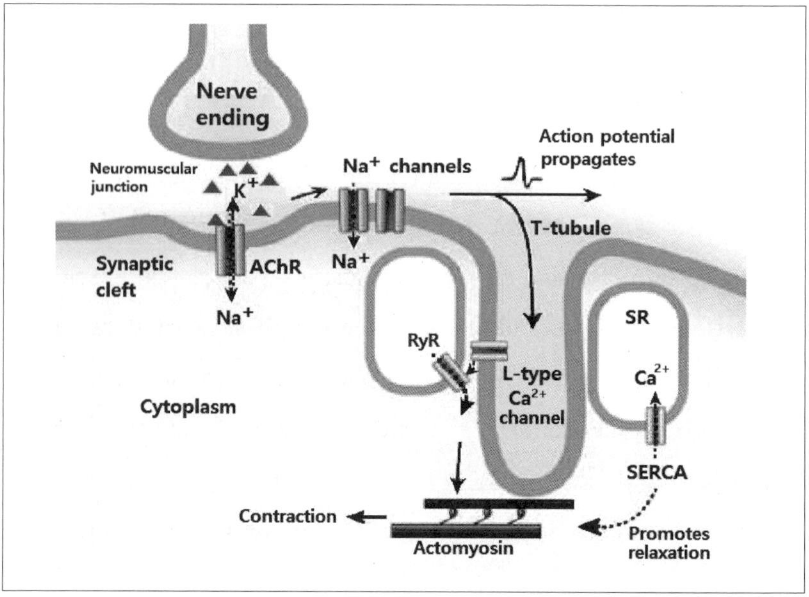

Figure 4.9. Skeletal muscle contraction and changes with exercise. Neurotransmitter (acetylcholine, ACh) released from nerve endings binds to receptors (AChRs) on the muscle surface. The ensuing depolarization causes sodium channels to open, which elicits an action potential that propagates along the cell. The action potential invades T-tubules and causes the L-type calcium channels to open, which in turn causes ryanodine receptors (RyRs) in the SR to open and release calcium, which stimulates contraction. Calcium is pumped back into the SR by (SR/ER calcium ATPase SERCA) pumps. The decreasing cytosolic calcium levels cause calcium to disassociate from troponin C and, consequently, tropomyosin reverts to a conformation that covers the myosin-binding sites.

The action potential runs along the top of the muscle and invades the T-tubules (specialized invaginations of the membrane containing numerous ion channels). The opening of voltage-gated sodium channels activates L-type voltage-gated calcium channels lining the T-tubule. A conformational change in these enables release of calcium on the closely apposed SR via activation of RyR1. Calcium then binds to troponin as described above, initiating the contraction process. Calcium-bound CaM also activates MLCK, whose phosphorylation of the MLC changes cross-bridge properties. This modulates the troponin- dependent contraction, although there is no effect on the ATPase activity of MLC. MLC phosphorylation instead enhances force development at submaximal saturating calcium concentrations. The phosphate group is subsequently removed by protein phosphatase 1 (PP1).

4. Contraction Mechanism

The cross-bridge cycle occurs in five steps (Figure 4.10). Initially, the myosin head is attached to an actin filament after the "power stroke"from the previous cycle and after the actomyosin complex has released adenosine diphosphate (ADP). In the absence of ATP, the system could remain in this rigid state for an indefinitely long period, as is indeed the case in rigor mortis. In this rigid state, the myosin head is at a 45-degree angle with respect to the actin and myosin filaments.

The ADP-free myosin complex ("attached state" in Figure 4.10) would quickly bind ATP at the concentrations of ATP normally found within cells. If unrestrained, this cross-bridge cycling would continue until depleting the cytoplasm of ATP. At that time, the muscle would remain in the stiff "attached state" because release of the cross-bridges from actin requires binding of ATP to myosin. As discussed earlier, muscle cells do not regulate cross-bridge cycling by modifying ATP. Instead, skeletal muscle and cardiac muscle control this cycle at the third step by preventing cross-bridge formation until the tropomyosin moves out of the way in response to an increase in Ca^{2+}.

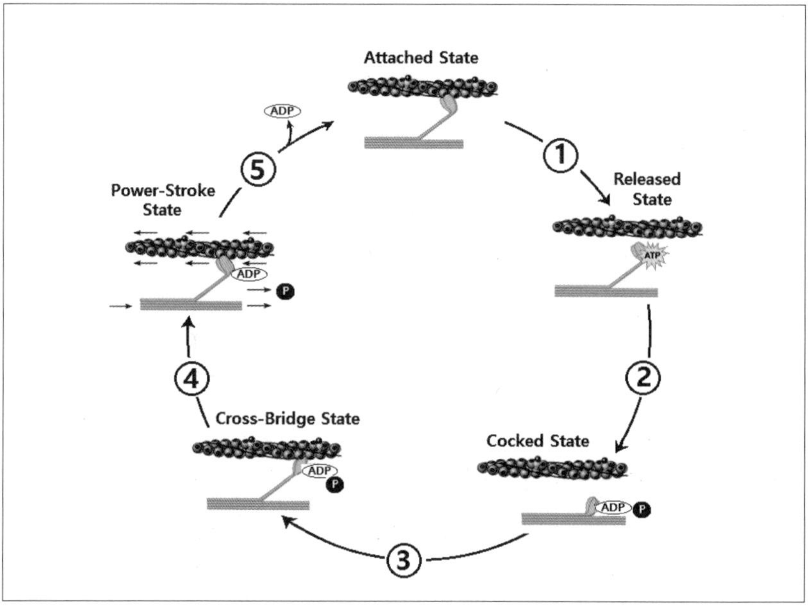

Figure 4.10. The cross-bridge cycle in skeletal and cardiac muscle. (1) ATP binds to myosin head, causing the dissociation of the actin-myosin complex. (2) ATP is hydrolyzed, causing myosin heads to return to their resting conformation. (3) A cross-bridge forms and the myosin nead binds to a new position on actin. (4) P is released. Myosin heads change conformation, resulting in the power stroke. The filaments slide past each other. (5) ADP is released.

Step 1: ATP binding. ATP binding to the head of the myosin heavy chain (MHC) reduces the affinity of myosin for actin, causing the myosin head to release from the actin filament. If all cross-bridges in a muscle were in this state, the muscle would be fully relaxed.

Step 2: ATP hydrolysis. The breakdown of ATP to ADP and inorganic phosphate (Pi) occurs on the myosin head; the products of hydrolysis are retained on the myosin. As a result of hydrolysis, the myosin head pivots around the hinge into a "cocked" position (perpendicular or at a 90-degree angle to the thick and thin filaments) This rotation causes the tip of the myosin to move about 11 nm along the actin filaments so that it now lines up with a new actin filament. Once again, if all cross-bridges in a muscle were in this state, the muscle would be fully relaxed.

Step 3: Cross-bridge formation. The cocked myosin head now binds to its new position on the actin filament. This binding reflects the increased affinity of the myosin-ADP-Pi complex for actin.

Step 4: Release of Pi from the myosin. Dissociation of Pi from the myosin head triggers the power stroke, a conformational change in which the myosin head bends approximately 45 degrees about the hinge and pulls the actin filament about 11 nm toward the tail of the myosin molecule. This conformational change causes the actin filament to be drawn along the myosin filament, thereby generating force and motion.

Step 5: ADP release. Dissociation of ADP from myosin completes the cycle, and the actomyosin complex is left in a rigid state. The myosin head remains in the same position and at a 45-degree angle with respect to the thick and thin filaments. The ADP-free myosin complex remain bound to actin until another ATP binds and initiates another cycle.

During contraction, the length of actin and myosin filaments does not change. Rather, the filaments slide along each other, pulling the Z disks closer to the myosin filaments, thereby decreasing sarcomere length. Figure 4.11 illustrates a series of positions that probably occur when myosin heads proceed through a cross-bridge cycle.

The A band width is constant during all phases of muscle contraction, but the I band and H zone widths change. These widths are greatest when the muscle is stretched (Figure 4.12A), and decrease as the muscle shortens (Figure 4.12B). In severely contracted muscle, actin filaments meet, or even overlap, in the center of the A band, and the Z disks may abut on the ends of the myosin filaments (Figure 4.12C). Under these conditions, the H zone and I band are not discernible in electron micrographs.

Muscle contraction requires energy in addition to that normally consumed by the resting muscle. This energy is derived from ATP. The enzyme responsible for hydrolyzing ATP into adenosine diphosphate (ADP) and inorganic phosphate is called myosin ATPase and the ATPase activity resides in the head region of the myosin molecule. The ATP splitting activity of myosin ATPase is enhanced by Ca^{2+}. Thus, increased free Ca^{2+} concentration in the sarcoplasm promotes cross-bridge formation between

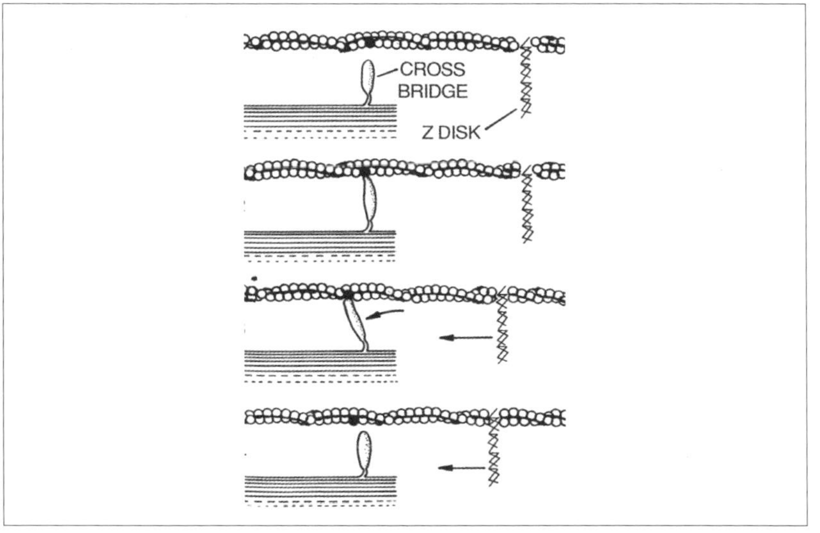

Figure 4.11. Sequence of positions of myosin heads during cross-bridge formation with actin.

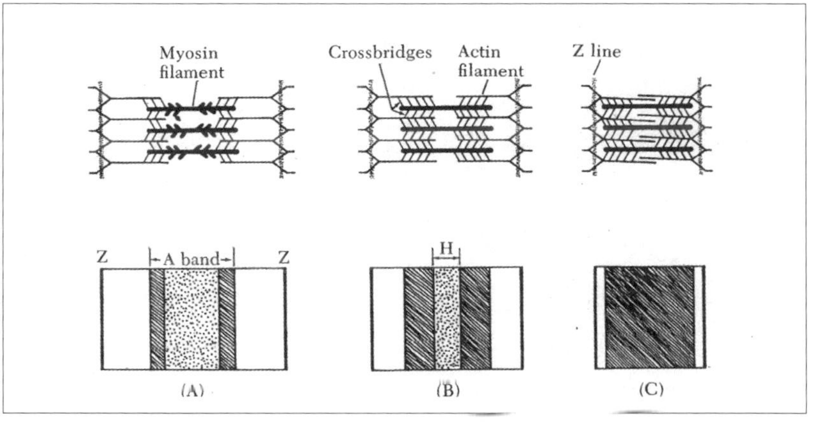

Figure 4.12. One sarcomere is shown at various stages of shortening during contraction (top). The banding pattern in a muscle fiber at corresponding degrees of contraction (bottom). Contraction is shown for the following stages: (A) in a muscle that is extended, (B) at rest length, and (C) severely shortened.

actin and myosin filaments, and simultaneously increases ATP-splitting activity of the myosin ATPase.

4. Terminating Contraction

In skeletal, cardiac, and smooth muscle, terminating contraction requires re-uptake of Ca^{2+} into the sarcoplasmic reticulum. After the contraction-activating stimulus has subsided, Ca^{2+} must be removed from the cytoplasm for contraction to cease and for relaxation to occur. Ca^{2+} may be extruded across the cell membrane or sequestered within intracellular compartments (Figure 4.13).

The cell may extrude Ca^{2+} using either an Na-Ca exchanger or a Ca^{2+} pump at the plasma membrane. Extrusion across the cell membrane, however, would eventually deplete the cell of Ca^{2+} and is therefore a minor mechanism for Ca^{2+} removal from the cytoplasm. Instead, Ca^{2+} re-uptake into the SR is the most important mechanism by which the cell returns Ca^{2+} to resting levels. Ca^{2+} re-uptake by the SR is mediated by a SERCA-type Ca^{2+} pump.

It is possible that the rate of Ca^{2+} re-uptake into the SR may be regulated by modulating the activity of the SR Ca^{2+} pump. For example, in cardiac muscle, SR Ca^{2+} pump activity is inhibited by the regulatory protein phospholamban. When phospholamban is phosphorylated by cyclic adenosine monophosphate (cAMP)-dependent protein kinase (PKA), its ability to inhibit the SR Ca^{2+} pump is lost. Thus, activators of

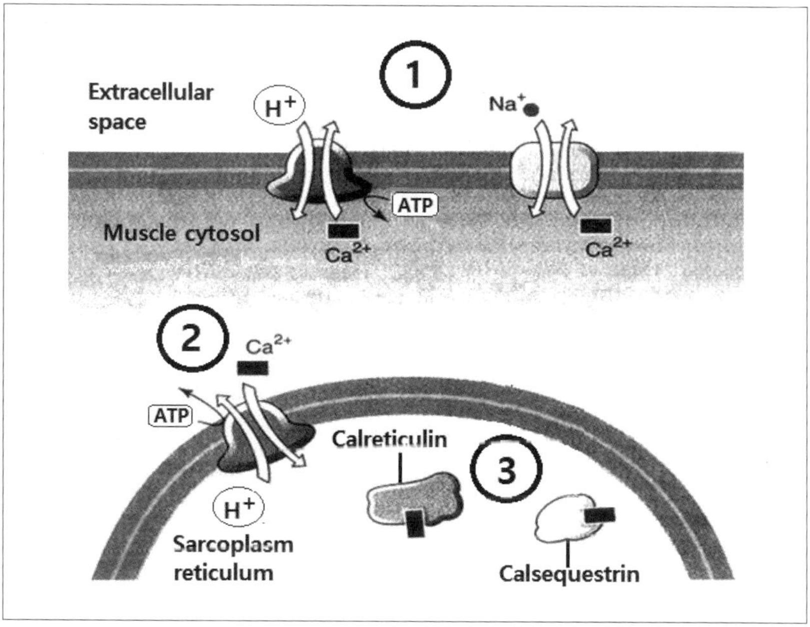

Figure 4.13. Mechanism of Ca^{2+} removal from the cytoplasm. (1) Na-Ca and exchanger and Ca^{2+} pump in the plasma membrane both extrude Ca^{2+} from the cell. (2) Ca^{2+} pump sequesters Ca^{2+} within the sarcoplasmic reticulum. (3) Ca^{2+} is bound in the sarcoplasmic reticulum by calreticulin and calsequestrin.

PKA, such as the neurotransmitter epinephrine, may enhance the rate of cardiac myocyte relaxation.

SR Ca^{2+} pump activity is also inhibited by high Ca^{2+} within the SR lumen. This inhibition of SR Ca^{2+} pump activity is delayed by Ca^{2+}-binding proteins within the SR lumen. These Ca^{2+}-binding proteins buffer the Ca^{2+} increase in the SR during Ca^{2+} re-uptake and thus markedly increase the Ca^{2+} capacity of the SR. The principal Ca^{2+}-

binding protein in skeletal muscle, calsequestrin, is also present in cardiac and some smooth muscle. Calreticulin is a ubiquitous Ca^{2+} -binding protein that is found in particularly high concentrations within the SR of smooth muscle. These proteins have a tremendous capacity to bind Ca^{2+}, with up to 50 binding sites per protein molecule.

Ca^{2+}-binding proteins are not located diffusely within the SR. Rather, calsequestrin is highly localized to the region of the SR immediately beneath the triad junction. Calsequestrin appears to bind directly at the triad, where it forms a complex with the Ca^{2+}-release channel and with two other triad proteins, junctin and triadin. Here, calsequestrin is poised not only to aid muscle relaxation by buffering Ca^{2+} within the SR lumen but also to unload its Ca^{2+} in the vicinity of the Ca^{2+}-release channel and thus facilitate excitation-contraction (EC) coupling. It has been hypothesized that EC coupling promotes Ca^{2+} release from calsequestrin, making Ca^{2+} available for exit from the SR.

Chapter 5.

Conversion of Muscle to Meat

Skeletal muscle tissue serves the function of locomotion in animals. This tissue also serves as an important source of food for man and is referred to as meat in this chapter. Muscles do not suddenly terminate all their living functions and become meat. The processes of change between muscles and meat are complex and involve metabolic, physical, and structural alterations. These changes occur over a period of several hours or even days after slaughter. At the death of an animal, the blood flow to and from the muscle ceases and as a result, there are no new sources of energy for muscle function and the supply of oxygen is cut off. In addition, the products of metabolism can no longer be removed and thus accumulated in the tissue. This process is referred to as the conversion of muscle to meat. This conversion of muscle to meat can be thought of as a gradual degradative process that, if it were allowed to continue indefinitely, would lead to a complete breakdown of muscle tissue to its constituent elements.

In the beginning of this chapter, it should be emphasized that all muscles and all parts of a muscle do not change in a uniform manner postmortem because of two major reasons. First, muscle consists of two major fiber types, red and white, containing different amounts of myoglobin, oxidative and glycolytic enzymes, and substrates. Different muscles contain different ratios of white to red fibers, and some muscles have major regions of predominantly white and/or red fiber. As a result, it can be easily expect a heterogeneous pattern of postmortem change even in adjacent cells, and adjacent fibers have different glycogen depletion patterns. Second, the effect of temperature on metabolic processes is profound and the rate of muscle cooling postmortem is not uniform, even within the same muscle. It is clear that a temperature gradient exists between the surface and the interior of the muscle. Also, the rate of cooling is affected noticeably by the size of the carcass, the temperature of the cooler, the rate of air movement past the carcasses, and the thickness of the insulating fat layer over the surface. Therefore, generalities about the rate of postmortem processes must be given with some caution in relation to muscle characteristics and temperature effects.

1. General Changes

1.1. Loss of Homeostasis

An understanding of the living function of muscle tissue is basic to understanding of the postmortem changes. The maintenance of a physiologically balanced internal environment is termed homeostasis.

This is system of checks and balances that provides the body with a means of coping with the stresses that tend to alter the internal environment. Homeostatic regulation gives an organism the ability to survive under many different, sometimes adverse, environmental conditions that may include extreme variations in temperature, oxygen deficiency, and psychological stress.

With increasing time postmortem, all homeostatic mechanism are eventually lost. As stored metabolites are depleted, muscle temperature declines because there are no operable mechanism for generating heat. Nervous control from the central system is lost within 4 to 6 minutes after exsanguination. Uncontrolled nerve impulses may arise locally, causing muscle twitches for some time after exsanguination. These local impulses may affect postmortem metabolism.

The altering of homeostasis is very important during conversion of muscle to meat for two reasons. First, many of the reactions and changes that occur during the conversion are a direct result of homeostasis. Second, conditions immediately prior to slaughter may alter postmortem changes and affect meat quality. This includes transportation of the animal from farm to slaughter house, its handling during the marketing process and its stunning or immobilization prior to slaughtering.

The first step of the harvesting (slaughtering) process is to immobilize animals. that is, to render them unconscious. Of course, the process must be done as quickly and efficiently as possible to minimize animal suffering. But, in addition, methods must be avoided that may cause tissue damage through elevations in blood pressure or excessive nerve stimulation. Extensive struggling and excessive movement, even though the animals are unconscious, often result in hemorrhage of blood into muscle or other tissues. In most developed countries, laws require humane immobilization methods that minimize animal distress and tissue damage. In Korea, captive bolt stunning is used for cattle, calves, and sheep; electrical stunning for calves, sheep, swine, poultry, and fish; and carbon dioxide immobilization for swine.

1.2. Circulatory Failure to the Muscle

Exsanguination, removal of blood, is the beginning of a series of postmortem changes in the muscles. It is not difficult to imagine that when the body senses massive bleeding, animal homeostasis is compromised. As soon as the homeostatic control mechanisms perceive a drop in blood pressure, the heart begins pumping faster and peripheral vessels constrict in an attempt to maintain pressure and force blood to the vital organs. As a result, only about 50 percent of the total blood volume can be removed from the body via the bleeding process while the remainder is held in the vital organs. Because blood is an excellent medium for growth of spoilage organism and excess blood in meat cuts is unappealing to the consumer, a thorough bleeding is essential to the slaughtering process.

As a result of exsanguination, the function of the circulatory system is lost. The circulatory system accounts for transportation of nutrients to the muscles and carrying metabolites and waste products away, either for excretion from the body or for further metabolism in other organs. After exsanguination, first of all, the ability of the oxygen supply to the muscle is lost. Muscle store only a limited amount of this important metabolites bound to the pigment myoglobin. In the living animal, oxygen is picked up in the lungs and carried to the cells of the body by the hemoglobin pigment of the blood. The myoglobin of the muscle cell has a greater attraction for oxygen than does hemoglobin, a characteristics that helps transfer oxygen from the blood to the muscle cells. The myoglobin provides a place for storing the oxygen until it is used by the cells for metabolism. The amount of oxygen stored in this manner is only sufficient to support oxidative reactions for a short period of time.

As the stored oxygen supply becomes depleted after exsanguination the aerobic pathway through the citrate cycle and the cytochrome system tend to stop functioning. Energy metabolism is then shifted to the anaerobic pathway, in much the same fashion as when there is insufficient oxygen for the living muscle during periods of heavy exercise and oxygen debt. Thus, muscle encounter another homeostatic mechanism in which an alternative source of energy is available to the muscles. It is clear that considerably less energy in the form of ATP is produced through the anaerobic path. However, the muscle is provided with a source of energy that will maintain the structural integrity and temperature of the cells for a while longer.

1.3. Postmortem Heat production

As a result of exsanguination, an important temperature control mechanism in muscle, the circulatory system, is lost. No longer can heat from deep part of the body be carried rapidly to the lungs and other surface areas for dissipation. Therefore, heat generated by ongoing metabolism causes a rise in muscle temperature soon after exsanguination. The extent of temperature rise depends on the rate of metabolic heat production and on its duration.

Some of the variation in heat production and dissipation encountered in muscle that may be accounted for by metabolic differences. Rapid glycolysis, as indicated by rapid pH decline, produces great quantities of heat, which results in slower carcass chilling. The size and location of muscles within the animal body, the amount of fat insulation, and predominate muscle fiber type (white vs. red) also influence the ultimate rise in temperature and rate of heat dissipation. It is important to recognize that the metabolic factors causing temperature to rise in postmortem muscle are the same ones that cause pH to decline. Therefore, protein denaturation becomes inevitable unless some means is employed to remove heat rapidly from muscle as it is produced.

External factors associated with the slaughter process also will influence heat dissipation. Scalding and singeing processes to which some carcasses are subjected contribute to a delay in heat dissipation. The ambient temperature in the slaughter room, duration of the slaughter and dressing operation, and temperature of the initial chill cooler all have considerable influence on the rate of carcass temperature decline.

1.4. Loss of Protective Mechanism

In healthy living animals, muscle is protected from invasions of microorganisms by a series of defenses, the first line of which is the tissues that cover the body and surround many of the internal organs. Connective tissues and cell membranes also may offer some protection. The lymphatic system and circulating white blood cells are available to destroy organism that enter the body.

During the conversion of muscle to meat, membrane properties are altered and the muscle becomes susceptible to bacterial invasion. Since the circulatory and lymphatic systems are no longer operating, they cannot prevent the spread of microorganism. Most of the changes that occur during conversion of muscle to meat favor microorganism proliferation. However, the lowering of muscle pH will have an inhibitory effect on some microorganism. Extra care must be exercised to prevent contamination of meat with spoilage organism during all postmortem handling and storage.

2. Postmortem Metabolism

2.1. ATPase Enzyme Systems

In living muscle, the complete oxidation of carbohydrate to carbon dioxide and water requires oxygen, and it releases a lot of energy. Much of this energy is captured by adding a phosphate group to another molecule that already has two phosphate groups. Thus, adenosine diphosphate (ADP) is converted to adenosine triphosphate (ATP). The ATP molecule carries this energy within the muscle fiber, and it may be released to another biochemical system by cleaving off the added phosphate (ATP→ ADP + Pi). Muscle contraction is a primary user of ATP in the living animal, but substantial amounts of ATP are used by the membranes around and within the fiber for maintaining ionic concentration gradients. One molecule of a carbohydrate may be combined with oxygen to add a phosphate group to each of 36 molecules of ADP. In addition to 36 molecules of ATP, this produces carbon dioxide and water. However, if the muscle is without oxygen (anoxic) after slaughter, the most it can normally gain by oxidizing each carbohydrate molecule is two molecules of ATP.

ATP is the major high-energy compound in cells and is used to drive a large variety of metabolic reactions including ion transport. Its presence is absolutely essential for cell survival. Thus, the postmortem muscle cell attempts to maintain its ATP level at all costs. In order to understand postmortem processes in muscle, the potential reactions and enzyme system in which ATP is split must be considered.

The enzyme possessing the highest potential ATPase activity in muscle is the myosin ATPase. In normal resting muscle, the myosin heads are not bound to actin and the million molar levels of magnesium present strongly inhibit the rate of ATP splitting. However, there is a continual slow hydrolysis of ATP to ADP and phosphate. Because the concentration of myosin is so large, this resting ATPase activity is probably the major ATP depletion reaction postmortem. The effect of temperature on ATP turnover in postmortem muscle closely parallels the temperature dependence of myosin ATPase at temperatures between 38 and 25℃.

The myosin ATPase activity is rapidly increased at least 300 fold if the sarcoplasmic calcium concentration is increased from levels 10-7 to 10-6 or 10-5 M. In living muscle, this calcium increase occurs during muscle contraction and cause myosin to combine with actin to form a complex whose ATPase activity is stimulated by magnesium. Consequently, the calcium level in sarcoplasm has a dramatic effect on postmortem metabolism, especially when visible contraction occurs.

The other ATPase system that must contribute to postmortem ATP degradation is the sarcoplasmic reticulum Ca-ATPase. This protein, which is embeded in the membrane, functions to move calcium out of the cytosol during the rest cycle in an ATP dependent transport process. Also, the mitochondrial ATPase must be consider for ATP splitting system. When mitochondria become anaerobic, they hydrolyze

ATP instead of synthesizing it. The specific activity of this enzyme is very high at temperature over 30℃ and could account for a major portion of the ATP turnover. However, the temperature dependence of the mitochondrial ATPase is much greater than that of ATP turnover in intact muscle, so it probably plays only a minor role.

2.2. Changes in Creatine Phosphate and ATP

Another high-energy compound is creatine phosphate (CP). CP is present in marked concentrations and serves to buffer ATP levels by participating in the following reactions: CP + MgADP → MgATP + Cr. Thus, even though the ATP splitting rate is rapid, the ATP concentration in muscle cell remains stable as long as this reaction occurs in the cell. In postmortem muscle, the ATP levels remain more or less constant until ~70% of the CP are degraded. At this point the ATP levels begin to decline rapidly and ultimately fall to almost zero.

The CP concentration in muscle removed from carcass within 10 min after death is already much lower than that in resting muscle. It is difficult to obtain resting values for CP and ATP. The total creatine in all species of meat animals is approximately 30~45 μmol/g and of this 13-23 μmol/g is in the phosphorylated form. The initial levels of CP in nonanesthetized muscles vary considerably with the species and the particular muscle examined. The decline in CP levels in beef longissimus muscles is reasonably slow with 1.5 μmol/g still remaining after 12 hr postmortem.

ATP levels normally remain fairly constant until the CP drops below 4 μmol/g. At 2 μmol/g CP, the ATP level is usually dropped 50%. Resting levels of ATP vary between 5.0 and 8.5 μmol/g. The ATP in the deeper portion of the semimembranosus may be virtually depleted by 5 hr postmortem. However, if this muscle is removed from the carcass and held at 0℃, the ATP depletion rate is drastically retarded with at least 12 hr required to lower ATP levels to <20% of the initial values.

Although there is wide variation in the rate of ATP depletion between muscles and species, the proportion of the original ATP concentrations at specific postmortem pHs is rather constant. Approximately one-eight of the ATP has been broken down by the time the muscle pH has dropped to pH 6.3, one-half is depleted by pH 5.9~6.0, and 90% is gone when the pH is 5.6~5.7.

2.3. Nucleotide Breakdown Products

ATP is split into ADP and inorganic phosphate by the ATPase. The ADP is usually rephosphorylated to ATP with CP or by glycolysis. However, with increasing time postmortem the CP is depleted and glycolysis can not keep up with ATP resynthesis. As a result, the following reaction occurs in an attempt to restore the resting ATP concentrations.

$$2ADP \xleftrightarrow{\text{myokinase}} ATP + AMP$$

The AMP is then deaminated to inosine monophosphate (IMP) by the following reaction.

$$\text{AMP} \xrightarrow{\text{ATP deaminase}} \text{IMP} + \text{NH}_3$$

The IMP is further degraded by the pathway below.

$$\text{IMP} \rightarrow \text{I} + \text{PO}_4 \xrightarrow{\text{nucleotide}} \text{hypoxanthine} + \text{ribose-1-phosphate}$$
$$\hookrightarrow \text{hypoxanthine} + \text{ribose}$$

The ADP levels in resting muscle are approximately 1.0 μmol/g. Much of this ADP is in a bound form, either complexed with actin or on the myosin heads. The actin ADP is tightly bound and accounts for the major portion of the remaining muscle ADP content at 1 to 2 days postmortem. The myosin ADP continues to cycle and will no longer be present when the ATP is fully depleted. In porcine muscles the ADP levels begin to decline within 1 hr postmortem while in bovine muscles the decline does not start until 6 hr after slaughter.

The concentration of AMP in resting bovine muscle is low. This level increases two to three times during the first 10 hr postmortem but declines to the original level by 24 hr postmortem. Inosine nucleotide levels in resting muscle are approximately 0.5 μmol/g, and most of this is IMP. The concentration of this compound increases postmortem in almost direct correspondence with the loss of ATP. The IMP levels increases to about 5.0 μmol/g between 12 and 24 hr postmortem in bovine muscles while similar levels are reached within 3 hr in porcine muscles.

The inosine and hypoxanthine levels are normally near zero in resting muscle. The appearance of these compounds depends on the IMP levels, and thus their production occurs primarily after glycolysis ceases. The production of hypoxanthine reaches 20~40% of the original ATP concentration by 24 hr postmortem in bovine muscle, and the level increases very slowly thereafter to a maximum after 60 days of cold storage at 2℃.

The concentrations of the nicotinamide dinucleotides (NAD) also change during postmortem. Resting bovine muscles contain 0.5~1.0 μmol/g, and these levels gradually decline to ~35% of their initial value by 8 hr postmortem and remain at this level through 36 hr postmortem. The NAD concentrations in porcine muscles decline by about one-third in the first 3 hr after slaughter. Lamb muscle NAD also declines to ~50% of initial levels by 24 to 48 hr postmortem. The degradation of NAD is markedly accelerated in muscle minces and homogenates.

2.4. Glycolysis

Skeletal muscle may use fatty acids, acetoacetate, or amino acids to obtain energy, but only the anaerobic pathway by which glycogen is converted to lactic acid. Glycogen is the primary storage carbohydrate in muscle fibers and appears in electron micrographs as single granules or clumps of granules located in the sarcoplasm between myofibrils and under the cell membrane. Glycogen is a polysaccharide formed by the linking together of large numbers of glucose units. The concentration of glycogen in fetal muscle increases steadily towards the end of gestation and may peak at the time of birth. Pigs have very high muscle glycogen levels at birth but they decline rapidly after birth to reach future adult levels within a week (Figure 5.1).

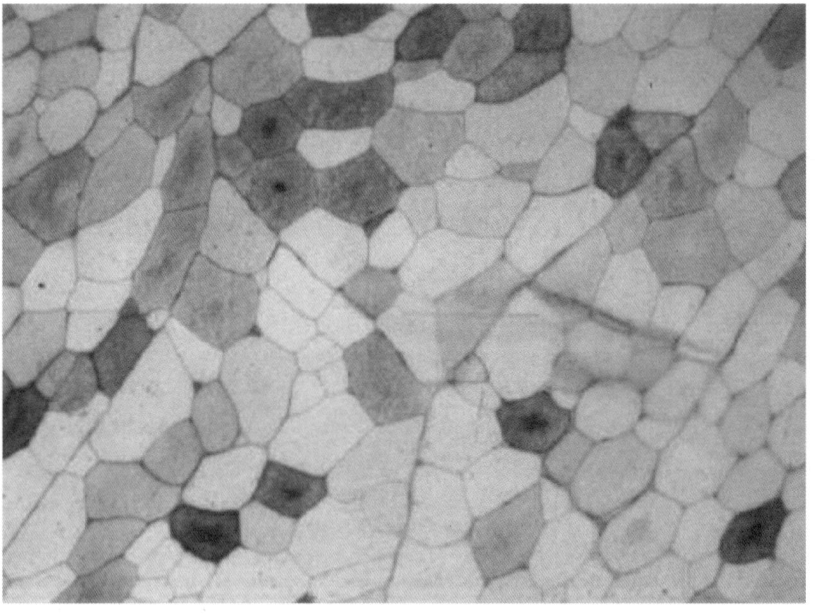

Figure 5.1. A transverse section of porcine muscle, just as most of its glycogen (stained dark) has been depleted.

There is currently a lot of interest in the possibility of postmortem glycogen break-down (glycolysis) by pathways other than the normal pathway mentioned above. Glycolysis is the enzymatic degradation of glycogen, and is the first step in the release of energy by the oxidation of glucose units. In life ATP is replenished by either oxidative metabolism in the mitochondria or by anaerobic glycolysis. When the blood supply ceases, there is no longer a source of oxygen and the mitochondrial ATP production is halted. No new glucose can be transported to the cell and thus glycolysis of the glycogen stores is the only way that ATP can be resynthesized. The end product of these reactions is lactic acid, which increases for some time postmortem. As a result, the muscle pH drops from 7.1~7.3 to 5.5~5.7 range.

A key point is why the conversion of muscle to meat is accompanied by the production of lactate; variability in lactic acid production is a major factor in meat quality. While the structure of glycogen and the mechanism of glycolysis are important in understanding the conversion of muscles to meat, the remainder of the pathway by which glycogen is anaerobically oxidized to lactate need only be covered in general principle. The most important question to be answered is why lactate is produced under anaerobic conditions, yet hardly at all under aerobic conditions?

After a series of steps in the glycolytic pathway, molecules with six carbon atoms derived from the glucose units of glycogen are split to produce two molecules of pyruvate, each with three carbon atoms. All the glycolytic enzymes (except for hexokinase) are concentrated in the I band. If aerobic conditions prevail, pyruvate formed in the cytosol of the muscle fiber now enters a mitochondrion (the metabolic furnace of the cell). After entering a mitochondrion, pyruvate is converted to acetyl-CoA which becomes fused to oxaloacetate to form citrate. The citrate then is oxidized in the well known Krebs' cycle, which is completed by the regeneration of oxaloacetate. Continuous activity of the Krebs' cycle is fuelled by a range of carbohydrates, fatty acids and amino acids, and is the primary system for the aerobic generation of energy. Large numbers of molecules of ATP are produced from ADP by a series of reactions, oxidative phosphorylation, that occur in the mitochondrial membrane.

Under aerobic conditions, the production of two pyruvate molecules from a glucose-1-phosphate molecule results in the reduction of 2NAD+. Thus, somewhere else in the muscle fiber, NADH must be re-oxidized for glycolysis to continue. Aerobically, this occurs as a consequence of mitochondrial Krebs' cycle activity, although NADH and NAD+ do not actually cross the mitochondrial membrane. In anaerobic living muscles and in meat, the Krebs' cycle is halted, and NADH is re-oxidized in the cytosol by lactate dehydrogenase (LDH) during the conversion of pyruvate to lactate. Pyruvate is of no immediate use anaerobically since mitochondrial oxidation has ceased, but its conversion to lactate ensures a continued supply of NAD+ for the continuation of anaerobic glycolysis in the cytosol. However, since these events form only the initial stages of complete carbohydrate oxidation, they do not regenerate much ATP. The net gain of ATP is reduced to only two molecules of ATP per molecule of glucose-1-phosphate. Molecules of glucose released from glycogen branch points generate a total net gain of 3ATP. LDH adds hydrogen to pyruvate to produce lactic acid.

The rate of glycolysis is linked directly to the rate of ATP splitting in muscle. Glycolysis requires ADP and phosphate to proceed, and the system stops when these compounds are converted to ATP. The glycolytic rate increases in direct proportion to the amount of ATPase activity added to the system. The concentration and activity of the glycolytic enzymes is sufficient to synthesize ATP at the rate of 100 μmol/g per min. This contrasts with typical ATP turnover rates of 0.50 to 0.65 μmol/g per min in normal muscle postmortem and maximal rates of approximately 10 μmol/g per min in extremely fast glycolyzing muscle.

The metabolic state of the fiber is profoundly affected by hormones, particularly adrenaline, and by the extent of recent contractile activity of the fiber. Phosphorylase is particularly important in the conversion of muscle to meat since it may be a primary control site for postmortem glycolysis. Because phosphorylase is abundant in fast contracting myofibers, there are complex relationships between myofiber contraction speed and rate of lactate production in meat. The factor that is most closely related to the ultimate pH (pHu) is the proportion of phosphorylase in the a form. Most phosphorylase in muscle is in the b form, which has less than 0.1% of phosphorylase a's activity. Phosphorylase b is also strongly inhibited by ATP. The other factor that affected pHu is the AMP deaminase activity. The activity of this enzyme is maximal at pH 6.2, and therefore a competition exists for nucleotide, with the result that total adenine nucleotides are progressively and irreversibly converted to IMP. As a result glycolysis ceases because there is no longer any ADP available for rephosphorylation.

3. Physico-chemical Changes

3.1. Muscle pH Decline

The lactic acid produced by anaerobic metabolism in the living animal is transported from the muscles to the liver, where it is re-synthesized into glucose and glycogen, or to the heart, where it is metabolized to carbon dioxide and water. However, during conversion muscle to meat, lactic acid remains in the muscle and increases in concentration during the postmortem period because the circulatory system is no longer capable of removing metabolites from the tissues of exsanguinated animals. The lactic acid continues to accumulate until the glycogen stored in the muscle is depleted.

Consequently, the muscle pH declines during postmortem as a result of the accumulation of lactic acid. Thus, the measurement of pH is a reliable and sensitive indicator of the rate and extent of postmortem glycolysis. The rate and extent of muscle pH decline after an animal has been exsanguinated are highly variable. In general, the initial pH values of resting muscles from the different species range from 7.1 to 7.4. These values for the bovine longissimus muscle decline normally to an ultimate level of 5.4 to 5.7.

A normal pH decline in porcine muscle is represented (Figure 5.2) by a gradual decrease from approximately pH 7.4 in living muscle to a pH of about 5.6 to 5.7 within 6 to 8 hours postmortem, and then to an pHu of about 5.3 to 5.7. In some animals the pH drops only slightly during the first hour after slaughter and then remains stable at a relatively high level, giving an ultimate pH in the range of 6.5 to 6.8. In

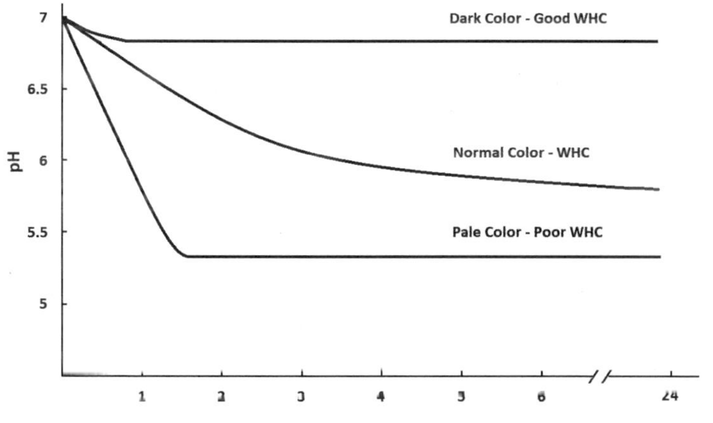

Figure 5.2. pH changes with time

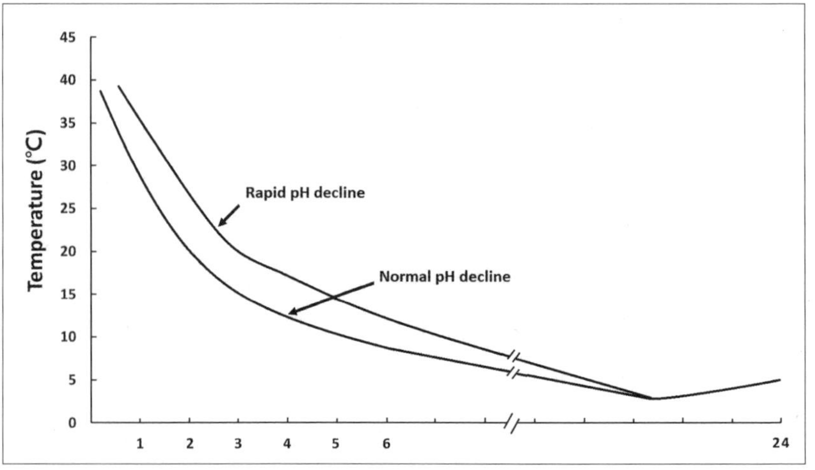

Figure 5.3. pH changes with temperature

other animals, muscle pH drops rapidly to around 5.4 to 5.5 during the first hour after exsanguination.

The rate of pH decline is markedly affected by temperature (Figure 5.3). The pH fall is rapid at high temperature in the range from 37℃ to 7℃. However, at temperatures <10℃ the rates plateau or even increase slightly. Also, the pH decline rate is species dependent because considerable variability exists between different animals in the rate of glycolysis even with the same rate of cooling.

The lowering of pH in muscle due to the accumulation of lactic acid is one of the most significant postmortem changes that can adversely affect meat quality. Development of acidic conditions in the muscle, before the natural body heat and the heat of continuing metabolism have been dissipated through carcass chilling, causes denaturation of the muscle proteins. The amount of denaturation depends upon how high temperature and low a level of pH is reached. As mentioned, temperature plays a key role in denaturation, and the denaturation of proteins causes a loss of protein solubility, loss of water-holding capacity, and loss in intensity of muscle pigment coloration. All these changes are undesirable, whether the muscle is going to be used as fresh meat or subjected to further processing.

Muscles which have a very rapid and/or extensive pH decline (Figure 5.2) will be pale in color, and have very low water-holding capacity that will give a cut surface a very wet appearance. In severe cases, fluid actually drips from the surface of the meat. On the other hand, muscles that maintain a high pH during conversion of muscle to the exposed cut surface, because naturally occurring water is tightly bound to proteins. On the other hand, somewhat different pattern of pH decline occurs in the surface of muscles exposed to the air. Oxidative metabolism continues, results in retarding of the lactic acid buildup. The muscle pH at 3 days postmortem is often over pH 6.0 in lamb and beef muscle sites.

3.2. Rigor Mortis

The most evident change in muscle postmortem is stiffening of muscle during conversion of muscle to meat. Muscle transformation from being soft, pliable, and stretchable to a more rigid and inextensible state is called as rigor mortis (Latin for "stiffness of death"). This phenomenon of rigor mortis has been recognized for hundreds of years but only during the last several decades has its physiological basis been explained.

The stiffening observed in rigor mortis is due to formation of permanent cross-bridges in muscle between actin and myosin filaments. The loss of extensibility is directly related to the loss of ATP concentrations postmortem. The rigidity occurs because the myosin heads firmly attach to actin and thus no longer allow free sliding of thick filaments past the thin filaments. This is the same chemical reaction that forms actomyosin during muscle contraction in living muscle. The difference between living and rigor muscles is that relaxation is impossible in the latter state, because no energy is available for breaking actomyosin bonds.

The rigor mortis patterns can be divided into three phases: delay, rapid and post-rigor (Figure 5.4). The delay phase is that period of time in which there is no change in the muscle elasticity. During the period immediately following exsanguination, muscle is quite extensible. If the muscle is loaded, or if a force is applied, it will passively stretch; and, when the force is removed, the natural elasticity of the muscle will return in to its original length. At this time, few if any actomyosin bridges are present to prevent force-induced extension. The rapid phase begins when the muscle extensibility begins to decline and ends when the extensiblility reaches its minimum. As cross-bridges form, muscles contract, shortening sarcomere and increasing muscle tension. These events are maximal when the ATP supply has been exhausted and all binding sites have formed actomyosin bridges. In the absence of ATP, the actomyosin bonds cannot be broken; consequently, tension is at its maximum. The post-rigor phase refers to the period after the rapid phase is completed. During postmortem storage, tension decreases but never returns to pre-rigor levels. The decrease in tension with time postmortem is a misnomer "resolution" of rigor mortis. This is a misnomer because the actomyosin bonds of rigor mortis are not broken during postmortem storage as implied by the term "resolution" of rigor. Rather, the decrease in tension is due to events other than breaking actomyosin bonds. Specifically, proteolytic degradation of specific myofibrillar proteins results in dissolution of Z disks and loss of ultrastructural integrity.

The length of time of the various rigor phases varies with species, muscle, holding temperature, and extend of struggling at the time of slaughtering. There is much more variation in the length of the delay phase than in the rapid phase. The time to rigor completion, end of the rapid phase, is greater in bovine muscles than in porcine muscles. This means that muscles with greater white fiber proportion have a longer delay phase than those that have predominantly red fibers. Rigor mortis is completed more rapidly at higher temperatures in the range 37~15℃. Also, struggling at death markedly shortens the rigor process in meat animals.

Muscle will shorten if unrestrained as it goes into rigor or will

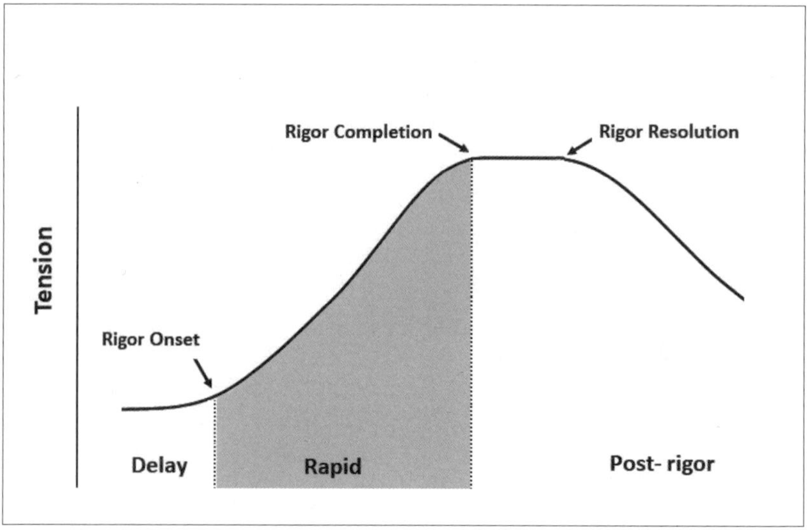

Figure 5.4. Different state of rigor mortis

develop tension if maintained at constant length. Temperature effects on beef muscle shortening and tension development is maximum at 1℃, smaller at 37℃, and very little at 16~25℃. Tension declines after rigor mortis completion, and this may explain the softening that occurs in muscle with increasing time post-rigor. At high temperatures the initiation of shortening under light loads is pH dependent. The extent of shortening at this pH depends on the remaining ATP concentration and the load on the muscle. This tendency to shorten during rigor is reflected in the decreased sarcomere lengths of post-rigor versus pre-rigor muscles. Maximum isometric tension occurs at the same time as the completion of rigor, and this rigor shortening can be prevented by antemortem injection of calcium chelators.

The pattern and cause of shortening that occurs at low temperature in muscle postmortem is quite different from that at high temperature. Bovine muscle removed from a beef carcass immediately postmortem is shortened by up to 50% when placed at 1℃. This shortening is called "cold shortening" and is distinct from rigor shortening because the former occurs more rapidly at high pH, and with ATP levels of 5 to 6 mM while the latter develops slowly at low pH, and at ATP levels <1mM. Cold shortening occurs only in muscles with a significant proportion of red muscle fibers and thus its extent is much less in most porcine muscles. Shortening occurs mainly as a result of a rise in intracellular calcium levels in the region of the myofibrils. This calcium is related by the sarcoplasmic reticulum because of its reduced efficiency in the cold or from mitochondria.

3.3. Loss of Structural Integrity

As mentioned above, pre-rigor muscle is soft, pliable, and highly extensible for several hours following slaughter. However, rigor mortis sets in within about 24 hours after slaughter and is accompanied by stiffening and hardening of the muscles. The muscle is shorten and lose their elasticity. At the onset of rigor, the muscle tissue no longer contracts on stimulation by electric current. As the muscle passes through rigor mortis, it loses some of the rigidity, but does not become soft and extensible again until autolysis reaches an advanced stage.

During development and completion of rigor mortis, the microscopic structure of muscle has much the same appearance as living muscle except sarcomere lengths have become shorter. However, it is almost certain that subtle degradative changes, such as loss of membrane integrity and function, begin soon after exsanguination. Carcasses and primal meat cuts are kept at refrigeration temperatures for various lengths of time after slaughter depending on their usage and merchandising schedules. During this postmortem period, many changes occur that alter meat properties. Most notable among the changes is an increase in tenderness due to the action of proteolytic enzymes.

The action of proteolytic enzymes in postmortem muscle is remarkably limited. Subtle changes occur in muscle proteins, as evidenced by the increasing degree of fragmentation that myofibrils undergo upon homogenization with time postmortem. The fragmentation is much greater in muscles before rigor mortis, and thus rigor tension is physically altering the myofibrils. In gel electrophoresis patterns with time postmortem, troponin-T is parally degraded and new bands in the 25,000~30,000 dalton region appear in postmortem muscles. There is also evidence for myosin degradation postmortem. Myosin heavy chain is split at several different points in postmortem beef muscle, especially after holding at 37℃. The fragments formed included components in the 50,000~100,000 dalton region, one of which appeared to be subfragment-1. Two myofibrillar proteins, connectin and desmin, are also degraded with time postmortem, and desmin content declines postmortem at about the same rate as troponin-T degradation. Since desmin is a Z disk protein, this protein degradation is related to the postmortem increase in fragility of myofibrils and ultrastructural changes in the Z disks.

Electron photomicrographs taken at increasing lengths of time postmortem show a progressive disruption of the myofibrillar structure (Figure 5.5). This degrative change causes muscle to become more extensive with increased storage time. Complete loss of these structures (Z disks) occurs due to proteolytic degradation of proteins associated with the disk, notably desmin and titin. While several other proteins of the myofibril are also degraded during postmortem storage, myosin and actin are not. The contribution of proteolytic degradation is still under investigation.

Summary of the proteolytic changes that occur in skeletal muscle during postmortem aging is as following: 1) Z disk degradation leads to weakening and fragmentation of myofibrils. 2) Degradation of desmin causes disruption of transverse cross-linking between myofibrilas and leads to fragmentation of myofibrils. 3) Degradation and disappearance of troponin-T. The exact relationship between meat tenderness and troponin-T degradation is not fully understood because of its location in myofibrils. 4) Degradation of titin and nebulin leads to fragmentation of myofibrils because of their ability to maintain longitudinal stability of myofibrils. 5) Degradation of above myofibrillar proteins results

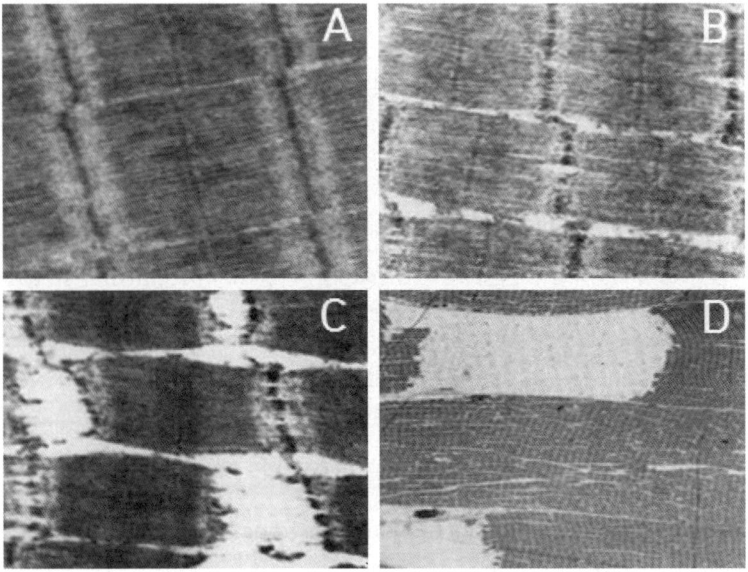

Figure 5.5. Degradation of Z disks in beef longissimus muscle during postmortem storage at 2℃. (A) One hour after slaughter, Z disks are intact (x 12,500). (B) Twenty-four hours after slaughter, some Z disk degradation is apparent (x 12,500). (C) Fourty-eight hours after slaughter, Z disk degradation and myofibril breaks are extensive (x 12,500). (D) Low magnification shows complete lateral breaks of he myofibrils in several muscle fibers 8 days after slaughter (x 650).

in appearance of new polypeptides of 95,000 and 28,000 to 32,000 molecular weight as seen by polyacrylamide gel electrophoresis. Perhaps the most significant observation is that myosin and actin, the major contractile proteins, are not affected even after 56 days of postmortem aging.

Several enzyme systems present in the postmortem proteolytic degradation of myofibrillar proteins. These systems included the calpains, the multicatalytic proteinase complex (both of which are present in the sarcoplasm), and the cathepsins, present in the lysosomes of muscle fibers. To qualify as the active participant, the enzyme(s) must degrade the same myofibrillar proteins that are degraded during postmortem storage of meat.

The multicatalytic proteinase complex does not cause postmortem proteolysis of any myofibrillar protein that plays a role in improving meat tenderness. Evidence indicates that the cathepsins are not released from lysosomes even after electrical stimulation and extended postmortem storage. Without their release, these enzymes cannot degrade myofibrillar proteins. The proteins would have to be endocytosed into lysosomes for proteolysis to occur as is the case in living muscle, but this is impossible in postmortem muscle because endocytosis is an active process requiring energy. Of these three enzyme system, the calpains produce all of the proteolytic changes in myofibrillar proteins during postmortem storage.

The calpain system consists of two calcium dependent enzymes, and a specific inhibitor, calpastatin. The two enzymes of this system differ in their calcium requirement for activation. One calpain requires milimolar calcium (m-calpain) and the other, micromolar calcium (μ-calpain) concentrations for activation. Calcium released from mitochondria and the sarcoplamic reticulum during postmortem storage activates the calpains.

Evidence is rather convincing that calpains are the enzymes accountable for the proteolytic changes in postmortem muscle. Infusion of calcium into carcasses or meat increases meat tenderness and produces all of the proteolytic changes. In contrast, when a chelator of calcium is infused to tie up calcium irons, none of the proteolytic changes occur and tenderness is not improved. Compared with the calpain system, however, neither the multicatalytic proteinase complex nor the cathepsins are affected by calcium.

The third component of the calpain system, calpastatin, which inhibits calpain activity, varies in amount in beef. Cattle whose meat tends to be "tough" (bulls or those with Bos indicus breeding, such as Brahman cattle) typically have a greater content of calpastatin in their muscles. Thus, high calpastatin content in these cattle limits the extent of postmortem proteolysis by the calpains, and consequently the tenderization process is reduced. Selection of cattle for low calpastatin potentially offers a means of improving tenderness of beef.

Improvement in tenderness during postmortem storage is due almost entirely to proteolytic degradation of myofibrillar proteins. Yet, degradation of collagen would dramatically improve meat tenderness, especially if intermolecular cross-linkages were degraded. However, postmortem proteolysis of collagen is minimal, and intermolecular cross-linkages are not degraded. The small amount of proteolytic degradation of collagen that does occur appears to take place only during the early postmortem period.

3.4. Changes in the Physical Appearance of the Muscle

In living animals, muscles with a sufficient oxygen supply has a bright red appearance. If the muscle is in oxygen debt, it is dark red or purple in color. In postmortem muscle, as the oxygen is used in metabolism, the muscle becomes dark purplish red in color. When fresh meat is cut, the exposed surface has this dark red appearance immediately after cutting. But, on exposure to the atmosphere for a few minutes, the myoglobin becomes oxygenated and changes to a brighter red color. If the muscle has been subjected to severe denaturation, the color intensity will be reduced to a pale red.

Living muscles maintain a certain amount of "tone" because skeletal attachments impose a small degree of tension or stretch. As muscles enter full rigor they become even firmer to the point of stiffness. Later in the conversion process, as enzymatic degradation and protein denaturation proceed, they become less firm. If protein denaturation is severe, muscles can become very soft.

Water accounts for 65 to 80 percent of total muscle mass. In living muscle fibers, as in other living cells, water plays a major role in cellular function. It acts as a solvent or carrier for substances that must be transported within the cell, and is an essential component of many chemical reactions. Much of the water in a muscle cells is tightly bound to various proteins. If proteins are not denatured, they continue to bind water during conversion of muscle to meat, and to a great extent, even during cooking. This retained water contributes to juiciness and palatability of meat as a food.

The changes that occur in water-binding during the conversion of muscle to meat, depend on the rate and extent of pH decline and on the amount of protein denaturation. When the pH of postmortem muscle remains very high (>6.5), the water-binding properties of meat are similar to those of living muscle. When the pH drops rapidly during the early postmortem period, a low water-binding capacity results.

Certain aspects of color, tenderness, juiciness, and flavor are influenced by the changes that occur during conversion of muscle to meat. Also, processed meat characteristics are affected by these changes. Thus it is obvious that postmortem changes are highly variable and influence the way in which muscle can best be utilized as meat. These changes can be controlled, to a certain extent, to improve meat products quality. Therefore, the antemortem and postmortem factors affecting this conversion are extremely important in the overall effort to control and improve the meat products quality.

References

Aberle, E. D., Forrest, J. C., Gerrard, D. E., and Mills, E. W. (2001). Principles of Meat Science. Kendall/Hunt Publishing Company.

Almenar-Queralt, A., Gregorio, C. C., and Fowler, V. M. (1999). Tropomodulin assembles early in myofibrillogenesis in chick skeletal muscle: evidence that thin filaments rearrange to form striated myofibrils. J. of Cell Science, 112: 1111-1123.

Bailey, A. J., and Light, N. D. (1989). Connective Tissues in Meat and Meat Products. New York, Elsevier Applied Science.

Bechtel, P. J. (1986). Muscle Development and Contractile Proteins. In Muscle as Food, P. J. Bechtel, ed. New York, Academic Press. pp. 1-35.

Bendall, J. R. (1971). Muscles, Molecules and Movement. New York, American Elsevier Publishing Company, Inc. pp. 3-57.

Bendall, J. R. (1973). Postmortem changes in muscle. In The structure and Function of Muscle, G. H. Bourne, ed. New York, Academic Press, pp. 244-309.

Bendall, J. R., and Swatland, H. J. (1988). A review of the relationships of pH with physical aspects of pork quality. Meat Science, 24: 85-126.

Briggs, F. N., Poland, J. L., and Solaro, R. J. (1977). Relative capabilities of sarcoplasmic reticulum in fast and slow mammalian skeletal muscles. J. of Physiology, 266: 587-594.

Casella, J. F., Craig, S. W., Maack, D. J., and Brown, A. E. (1987). CapZ(36/32), a barbed end actin-capping protein, is a component of the Z-line of skeletal muscle. J. of Cell Biology, 105: 371-379.

Cassens, R. G. (1971). Microscopic Structure of Animal Tissues. In The Science of Meat and Meat Products, J. F. Price and B. S. Schweigert, eds. San Francisco, W.H. Freeman and Company. pp.11-77.

Cassens, R. G. (1987). Structure of Muscle. In The Science of Meat and Meat Products, J. F. Price and B. S, Schweiger, eds. Westport, Ct, Food and Nutrition Press. pp. 11-59.

Grove, B. K., Holmbom, B., and Thornell, L. -E. (1987). Myomesin and M protein: differential expression in embryonic fibres during pectoral muscle development. Differentiation, 34: 106-114.

Huxley, H. E. (1965). The Mechanism of Muscular Contraction. Sci. Am. 213, 18. Offprint No. 1026. San Francisco, W.H. Freeman and Company.

Joo, S. T., and Kim G. D. (2011). Meat quality traits and control technologies. In Control of Meat Quality, S. T. Joo, ed. India, Research Signpost.

Joo, S. T., Kim, G. D., Hwang, Y. H., and Ryu, Y. C. (2013). Control of fresh meat quality through manipulation of muscle fiber characteristics. Meat Science, 95: 828-836.

Kauffman, R. G., and Clair, L. E. St. (1965). Porcine Myology. University of Illinois College of Agriculture, Agricultural Experiment Station, Bulletin 715.

Lawrie, R. A. (1979). Chemical and Biochemical Constitution of Muscle. In Meat Science, 3rd ed. New York, Pergamon Press. pp. 75-130.

Lawrie, R. A. (1985). Meat Science. 4th edition. Pergamon Press, Oxford.

Labeit, S., and Kolmerer, B. (1995). The complete primary structure of human nebulin and its correlation to muscle structure. J. of Molecular Bilogy, 248: 308-315.

Maruyama K., and Ebashi, S. (1970). Regulatory Proteins of Muscle. In The Physiology and Biochemistry of Muscle as a Food, II, E. J. Briskey, R. G. Cassens, and B. B. Marsh, eds. Madison, University of Wisconsin Press PP. 373.

MacLennan, D. H., and Phillips, M. S. (1992). Malignant Hyperthermia. Science 256, 789.

Maier, A. E., Leberer, E., and Pette, D. (1986). Distribution of sarcoplasmic reticulum Ca-ATPase and of calsequestrin in rabbit and rat skeletal muscle fibres. Histochemistry, 86: 63-69.

Marsh, B. B. and Carse, W. A. (1974). Meat tenderness and the sliding-filament hypothesis. J. of Food Technology. 9: 129-139.

Ohtsuki, I. (1999). Calcium ion regulation of muscle contraction: the regulatory role of troponin T. Molecular and Cellular Biochemistry, 190: 33-38.

Orcutt, M. W., Dutson. T. R., Wu, F. Y., and Smith, S. B. (1986). The fine structure of the endomysium, perimysium and intermyofibrillar connections in muscle. Food Microstructure, 5: 41-51.

Price, M. G. (1987). Skelemins: cytoskeletal proteins located at the periphery of M-discs in mammalian striated muscle. J. of Cell Biology, 104: 1325-1336.

Schroeter, J. P., Bretaudiere, J. P., Sass, R. L. and Golstein, M. A. (1996). Three-dimensional structure of the Z band in a normal mammalian skeletal muscle. J. of Cell biology, 133: 571-583.

Squire, J. (1981). The structural Basis of Muscular Contraction. New York, Plenum Press.

Swatland, H. J. (1984). Structure and Development of Meat Animals. Englewood Cliffs. NJ, Prentice-Hall.

Swatland, H. J. (1994). Structure and Development of Meat Animals. Technomic Publishing, Lancaster, Pennsylvania.

Weber, A., Pennise, C. R., Babcock, G. G., and Fowler, V. M. (1994). Tropmodulin caps the pointed ends of actin filaments. J. of Cell Biology, 127: 1627-1635.

Yamaguchi, M., Kamisoyama, H., Nada, S., Yamano, S., Izumimoto, M., Hitai, Y., Cassens, R. G., Nasu, H., Muguruma, M., and Fukazawa, T. (1986). Current concepts of muscle ultrastructure with emphasis on Z-line architecture. Food Microstructure, 5: 197-205.

Principles of

Muscle Biology

for Meat Science

초판 1쇄 발행 • 2018년 3월 5일

지은이 • 주선태
펴낸이 • 이상경
부　장 • 박현곤
편집장 • 김종길
디자인 • people communication

펴낸곳 • 경상대학교출판부
주소 • 경남 진주시 진주대로 501
전화 • 055) 772-0801(편집), 0802(디자인), 0803(도서 주문)
팩스 • 055) 772-0809
전자우편 • gspress@gnu.ac.kr
홈페이지 • http://gspress.gnu.ac.kr
페이스북 • https://www.facebook.com/gnupub
블 로 그 • https://gnubooks.tistory.com
등록 • 1989년 1월 7일 제16호

• 책값은 뒤표지에 있습니다.

이 도서의 국립중앙도서관 출판시도서목록(CIP)은 서지정보유통지원시스템 홈페이지(http://seoji.nl.go.kr)와 국가자료공동목록시스템(http://www.nl.go.kr/kolisnet)에서 이용하실 수 있습니다.
(CIP제어번호: CIP2018006438)